PASS THE MOT!

HOW TO GET YOUR CAR THROUGH THE ANNUAL TEST FIRST TIME

JIM PUNTER

Warning

While every attempt has been made throughout this book to emphasize the safety aspects of the various inspections when checking a vehicle, the publishers, the author and the distributors accept no liability whatsoever for any damage, injury or loss resulting from the use of this book. If you have any doubts about your ability to <u>safely</u> check your vehicle for MOT test purposes, then it is recommended that you desist from such action and have the vehicle inspected professionally.

The contents of this book relate to the MOT regulations at the time of writing. The publishers, the author and the distributors accept no liability whatsoever for errors, omissions or discrepancies arising from subsequent changes in the testing regulations.

First published in 1993

British Library Cataloguing-in-Publication Data:
A catalogue record for this book is available from the British Library.

ISBN 1-85010-861-7 (Book Trade)
ISBN 1-85010-864-1 (Motor Trade)

Published by Haynes Publishing, Sparkford, Nr Yeovil, Somerset BA22 7JJ.

Typeset by BPCC Techset Ltd, Exeter

Printed in Great Britain by J. H. Haynes & Co. Ltd

CONTENTS

Safety first!

Safety first!

However enthusiastic you may be about getting on with the job in hand, do take the time to ensure that your safety is not put at risk. A moment's lack of attention can result in an accident, as can failure to observe certain elementary precautions. There will always be new ways of having accidents, and the following points do not pretend to be a comprehensive list of all dangers; they are intended rather to make you aware of the risks and to encourage a safety-conscious approach to all work you carry out on your vehicle.

Essential DOs and DON'Ts

DON'T rely on a single jack when working underneath the vehicle. Always use reliable additional means of support, such as axle stands, securely placed under a structural part of the vehicle that will not give way.

DON'T attempt to loosen or tighten high-torque nuts (eg wheel hub nuts) while the vehicle is on a jack; it may be pulled off.

DON'T snatch backwards and forwards when checking suspension and steering components with the vehicle on jacks or ramps. It could topple off. Push and pull firmly but steadily and carefully.

DON'T start the engine without first ascertaining that the transmission is in neutral (or 'Park' where applicable) and the handbrake applied.

DON'T suddenly remove the filler cap from a hot cooling system—cover it with a cloth and release the pressure gradually first, or you may get scalded by escaping coolant.

DON'T attempt to drain oil, automatic transmission fluid, or coolant until you are sure it has cooled sufficiently to avoid scalding you.

DON'T grasp any part of the engine, exhaust or catalytic converter without first ascertaining that it is sufficiently cool to avoid burning you.

DON'T allow brake fluid or antifreeze to contact vehicle paintwork.

DON'T syphon toxic liquids such as fuel, brake fluid or antifreeze by mouth, or allow them to remain on your skin.

DON'T inhale dust—it may be injurious to health (see *Asbestos*).

DON'T allow any spilt oil or grease to remain on the floor—wipe it up straight away, before someone slips on it.

DON'T use ill-fitting spanners or other tools which may slip and cause injury.

DON'T attempt to lift a heavy component which may be beyond your capability—get assistance.

DON'T rush to finish a job, or take unverified short cuts.

DON'T allow children or animals in or around an unattended vehicle.

DON'T park vehicles with catalytic converters over combustible materials such as dry grass, oily rags, etc. if the engine has recently been run. As catalytic converters reach extremely high temperatures, any such materials in close proximity may ignite.

DON'T run vehicles equipped with catalytic converters without the exhaust system heat shields fitted.

DO wear eye protection when using power tools such as an electric drill, sander, bench grinder, etc., and when working under the vehicle.

DO use a barrier cream on your hands prior to undertaking dirty jobs—it will protect your skin from infection as well as making the dirt easier to remove afterwards; but make sure your hands aren't left slippery. Note that long term contact with used engine oil can be a health hazard.

DO keep loose clothing (cuffs, tie etc) and long hair well out of the way of moving mechanical parts.

DO remove rings, wristwatch etc, before working on the vehicle—especially the electrical system.

DO ensure that any lifting tackle or jacking equipment used has a safe working load rating adequate for the job, and is used precisely as recommended by the manufacturer.

DO keep your work area tidy—it is only too easy to fall over articles left lying around.

DO get someone to check

periodically that all is well when working alone on the vehicle.
DO carry out work in a logical sequence and check that everything is correctly assembled and tightened afterwards.
DO remember that your vehicle's safety affects that of yourself and others. If in doubt on any point, get specialist advice.
IF, in spite of following these precautions, you are unfortunate enough to injure yourself, seek medical attention as soon as possible.

Asbestos

Certain friction, insulating, sealing, and other products—such as brake linings, brake bands, clutch linings, gaskets, etc.—may contain asbestos. *Extreme care must be taken to avoid inhalation of dust from such products since it is hazardous to health*. If in doubt, assume that they *do* contain asbestos.

Fire

Remember at all times that petrol is highly flammable. Never smoke, or have any kind of naked flame around, when working on the vehicle. But the risk does not end there—a spark caused by an electrical short-circuit, by two metal surfaces contacting each other, by careless use of tools, or even by static electricity built up in your body under certain conditions, can ignite petrol vapour, which in a confined space is highly explosive. The vapour produced by spilling oil or hydraulic fluid onto hot metal, such as an exhaust manifold, can also be flammable or explosive.

Whenever possible disconnect the battery earth (negative) terminal before working on any part of the fuel or electrical system, and never risk spilling fuel on to a hot engine or exhaust. Catalytic converters run at extremely high temperatures, and consequently can be an additional fire hazard. Observe the precautions outlined elsewhere in this section.

It is recommended that a fire extinguisher of a type suitable for fuel and electrical fires is kept handy in the garage or workplace at all times. Ideally, a suitable extinguisher should also be carried in the vehicle. Never try to extinguish a fuel or electrical fire with water. If a vehicle fire does occur, take note of the remarks below about hydrofluoric acid.

Hydrofluoric acid

Hydrofluoric acid is extremely corrosive. It is formed when certain types of synthetic rubber, which may be found in O-rings, oil seals, brake hydraulic system seals, fuel hoses, etc., are exposed to temperatures above 400°C. The obvious circumstance in which this could happen on a vehicle is in the case of a fire. The rubber does not burn, but changes into a charred or sticky substance which contains the acid. *Once formed, the acid remains dangerous for years. If it gets onto the skin, it may be necessary to amputate the limb concerned*.

When dealing with a vehicle which has suffered a fire, or with components salvaged from such a vehicle, always wear protective gloves, and discard them carefully after use. Bear this in mind if obtaining components from a car breaker.

Fumes

Certain fumes are highly toxic, and can quickly cause unconsciousness and even death if inhaled to any extent, especially if inhalation takes place through a lighted cigarette or pipe. Petrol vapour comes into this category, as do the vapours from certain solvents such as trichloroethylene. Any draining or pouring of such volatile fluids should be done in a well-ventilated area.

When using cleaning fluids and solvents, read the instructions carefully. Never use materials from unmarked containers—they may give off poisonous vapours.

Never run the engine of a motor vehicle in an enclosed space such as a garage. Exhaust fumes contain carbon monoxide, which is extremely poisonous; if you need to run the engine, always do so in the open air, or at least have the rear of the vehicle outside the workplace. Although vehicles fitted with catalytic converters have greatly reduced toxic exhaust emissions, precautions should still be observed.

If you are fortunate enough to have the use of an inspection pit, never drain or pour petrol, and never run the engine, while the vehicle is standing over it; the fumes, being heavier than air, will concentrate in the pit, with possibly lethal results.

The battery

Batteries which are sealed for life require special precautions, which are normally outlined on a label attached to the battery. Such precautions are primarily related to situations involving battery charging and jump-starting from another vehicle.

Whatever the type of battery, never cause a spark, or allow a naked light, in close proximity to it. It will normally be giving off a certain amount of hydrogen gas, which is highly explosive.

Whenever possible, disconnect the battery earth (negative) terminal before working on the fuel or electrical systems.

If possible, loosen the filler plugs or cover when charging the battery from an external source. Do not charge at an excessive rate, or the battery may burst. Special care should be taken with the use of high charge-rate boost chargers to prevent the battery from overheating.

Take care when topping-up and when carrying the battery. The acid electrolyte, even when diluted is very corrosive, and should not be allowed to contact clothing, eyes or skin.

Always wear eye protection when cleaning the battery, to

prevent the caustic deposits from entering your eyes.

The vehicle electrical system

Take care when making alterations or repairs to the vehicle wiring. Electrical faults are the commonest cause of vehicle fires. Make sure that any accessories are wired correctly using an appropriately-rated fuse and wire of adequate current-carrying capacity. When possible, avoid the use of 'piggy-back' or self-splicing connectors to power additional electrical equipment from existing feeds; make up a new feed with its own fuse instead.

When considering the current which a new circuit will have to handle, do not overlook the switch, especially when planning to use an existing switch to control additional components—for instance, if spotlights are to be fed via the main lighting switch. For preference, a relay should be used to switch heavy currents. If in doubt, consult an auto electrical specialist.

Any wire which passes through a body panel or bulkhead must be protected from chafing with a grommet or similar device. A wire which is allowed to chafe bare against the bodywork will cause a short-circuit and possibly a fire.

Mains electricity and electrical equipment

When using an electric power tool, inspection light, diagnostic equipment, etc., which works from the mains, always ensure that the appliance is correctly connected to its plug and that, where necessary, it is properly earthed. Do not use such appliances in damp conditions and, again, beware of creating a spark or applying excessive heat in the vicinity of fuel or fuel vapour. Also ensure that the appliances meet the relevant national safety standards.

Ignition HT voltage

A severe electric shock can result from touching certain parts of the ignition system, such as the HT leads, when the engine is running or being cranked, particularly if components are damp or the insulation is defective. Where an electronic ignition system is fitted, the HT voltage is much higher and could prove fatal, especially to wearers of cardiac pacemakers.

Jacking and vehicle support

The jack provided with the vehicle is designed primarily for emergency wheel changing, and its use for servicing and overhaul work on the vehicle is best avoided. Instead, a more substantial workshop jack (trolley jack or similar) should be used. Whichever type is employed, it is essential that additional safety support is provided by means of axle stands designed for this purpose. Never use makeshift means such as wooden blocks or piles of house bricks, as these can easily topple or, in the case of bricks, disintegrate under the weight of the vehicle. Be particularly careful when the vehicle is either jacked up or on ramps, and an inspection has to be done which involves pushing and pulling by the assistant to detect excessive wear.

If removal of the wheels is not required, the use of drive-on ramps is recommended for inspections when the wheels have to remain under load. Caution should be exercised to ensure that they are correctly aligned with the wheels, and that the vehicle is not driven too far along them so that it promptly falls off the other ends, or tips the ramps. Always chock the wheels remaining on the ground.

Special precautions for diesel engines

Diesel injection pumps supply fuel at very high pressure. Extreme care must be taken when working on the fuel injectors and fuel pipes. It is advisable to place an absorbent cloth around the union before slackening a fuel pipe. *Never expose the hands, face or any other part of the body to injector spray: the high working pressure can penetrate the skin, with potentially fatal results.* Injector test rigs produce similarly high pressures and must be treated with the same respect.

Diesel fuel is more irritating to the skin than petrol. It is also harmful to the eyes. Used engine oil from diesel engines is more carcinogenic than that from petrol engines. Besides the use of a barrier cream to protect the hands, consider using lightweight disposable gloves when fuel or oil spillage is inevitable. Change out of fuel-soaked or oil-soaked clothing as soon as possible.

Spilt diesel fuel does not evaporate like petrol. Clear up spillages promptly to avoid accidents caused by slippery patches on the workshop floor. Note also that diesel attacks tarmac surfaces: if working at the roadside or in a drive, put down newspaper or a plastic sheet if fuel spillage is expected.

Fuel spillage

Whether the vehicle has a petrol or diesel engine, fuel spillage can be hazardous due to the risk of fire and inflammable or toxic fumes. Both petrol and diesel are corrosive, and can cause skin irritation with potential carcinogenic results. Both also cause damage to tarmac surfaces.

Always clear up spills immediately and safely dispose of any contaminated matter.

Acknowledgements

I would like to thank Peter
Jeffers for reading the manuscript
and offering valuable technical
advice from the perspective of
the DIY mechanic.

My thanks also to the
management and staff of Punters
Garages Ltd for their advice and
assistance throughout the
preparation of this book.

Introduction

Over 20 million vehicles are subjected to an MOT test every year, and unless your car is less than three years old it will need to be tested annually. For a lot of people 'the MOT' is an annual problem. Not only is there the inconvenience of having to take their car to the testing station, but, if it fails, there is then the cost and trouble of having it repaired and arranging yet another visit for the MOT re-test.

As the owner of a testing station I am surprised how little is known about the MOT test by members of the general public. So many questions are asked, such as: 'Do I need the registration document?' . . . 'Does the spare wheel have to be in good condition?' . . . 'Do you want the old MOT certificate?'

It is not just the inexperienced who are curious and, in many cases, unjustifiably anxious when taking their cars to be tested. I receive many calls from motorists who are DIY enthusiasts asking technical questions about the test before bringing their vehicle along, and it is just not always possible to answer these queries satisfactorily during a short telephone call.

So this book is to answer all those questions which people have asked about 'the MOT'. But it goes much further than that, and provides the DIY motorist with a comprehensive guide to and a complete insight into the MOT test for cars and light commercial vehicles. It shows how 'the MOT' is conducted, and what you can do to check your car before taking it into the testing station.

A vital companion to this book is the Haynes workshop manual for the vehicle in question. This is important because the design of cars and light commercial vehicles varies greatly from model to model, and it would be impossible in a single book to provide specific and detailed information for every make.

With the increasing use of scientific and sophisticated testing equipment, some aspects of the test cannot be fully duplicated by the DIY motor mechanic. But a lot *can* be done, and by carefully and diligently following the step-by-step guidance provided here, there will be much less chance of your car unexpectedly failing 'the MOT'.

Chapter 1

BEFORE YOU START

Cars, light vans and small trucks ('Class IV vehicles' in the official jargon) form well over 90 per cent of the vehicles to be found on the road which require a 'normal' MOT test. All have to be first tested when they are three years old, and annually thereafter. The vehicles covered in this book are in this category and include cars and light commercial vehicles up to 300 kg (3 ton) design gross weight, minibuses with up to 12 passenger seats and motor caravans.

In terms of motoring history the MOT test is a modern innovation, only having been introduced in 1961. Originally the test applied to all cars and light commercial vehicles which were over ten years old and was a simple examination of the brakes, lights and steering. Now it is much more thorough and comes into effect much sooner. The tested items are all checked in a particular way, sometimes using specialized equipment not readily available to the average motorist. Where this special equipment is needed it will be noted in the text.

The MOT test does not thoroughly check a car for roadworthiness. This causes a lot of confusion, and throughout this book it is vital to remember that:

The MOT is only an *examination* of certain parts of the car, carried out in a special way, using test equipment which has been approved by the authorities, and carried out by an authorized MOT tester.

This means that a pass certificate only applies *at the time of the test* and does not mean that the car is in good condition.

A car can pass the MOT but still be considered dangerous to use on the road because an item is faulty which is not part of the MOT test. A loose battery which could topple and spill acid is just one such example.

The moral here is to beware of the assumption, often encouraged by car dealers, that if the car has a new MOT certificate, it is in good condition—not necessarily so!

Safety precaution
When thoroughly examining any motor vehicle prior to submitting it for an MOT test, it will sometimes be necessary to either jack the vehicle up, or put it on a ramp to check things from underneath. At the same time parts of the vehicle will have to be pushed and pulled.

This part of the examination must be carried out very carefully with an eye to the security of the vehicle to ensure the safety of anybody beneath it. Such safety precautions are essential, and warnings are included at appropriate points in the book. Also see the safety notes on pages 6–8.

Finally, before you start, remember that many aspects of an MOT test depend to a great extent on the judgement of the tester, and he will often have to make borderline decisions which could go one way or the other. If he is faced with a very dirty vehicle which appears to have been neglected and not carefully maintained, then it is only human

Figure 1. The author's MOT garage at Hillingdon in West London.

nature that he is less likely to give the benefit of doubt in such cases. On the other hand, if the car is clean and tidy and seems to have been well looked after, it is fair to assume that he will be more likely to pass an item which he feels could go either way.

So, before taking your car to be tested, it is well worth while giving it a good clean.

Chapter 2

LIGHTING EQUIPMENT

The MOT test of your car's lighting equipment will depend to a great extent on the age of your car and the type of equipment originally fitted. Any variations will be considered in the following sections on the different lights and the way they are tested.

Of passing interest here is that if a vehicle is only used during daylight hours (whatever that may mean officially!) and does not have any lights fitted, or they are <u>permanently</u> disconnected, then they will not be subject to test. This applies to all the lights, including the direction indicators. In practice this rarely happens, so let's have a look at the normal requirement for testing a vehicle's lights.

Battery condition precaution
The examination of the lights can take some time and may flatten the battery unless the engine is running, so make sure an assistant holds the accelerator down slightly to make sure that the battery is being continually charged while the lights are being checked.

The lighting equipment on your car which will be tested is as follows:

Side lamps, known officially as front and rear position lamps.
Headlamps, but not including fog lamps and spot lamps.
Stop lamps.
Rear reflectors.

Direct indicators, including the semaphore type on older cars.
Hazard warning lamps, depending on your car's age.
Number plate lamps.
Fog lamps, depending on your car's age.

Taking these in turn:

SIDE LAMPS

What the MOT Test looks for
The aspects of the side lamps which have to be tested are:

* **Position, security and condition.**
* **Colour and brightness.**
* **Operation.**

How to check
Position security and condition: Front and rear side lamps must be positioned correctly and must not be loose. This means that they must be on each side of the vehicle, at about the same height and position, and face the right way. The tester checks this by eye. The lamps must not be incomplete, damaged, or faded to the extent that they cannot be seen from a reasonable distance. This, of course, will apply to all the lamps.

This is a simple check. If the car has its original lamps, then the size and position will not be an issue. Otherwise, check that the lamps are located about the same height, and positioned about the same distance either side of the vehicle centre line.

The lamps must not be loose or damaged or, as sometimes happens, have a lens which has 'aged' so that it dims the light. If the lamps are loose in their housing or the housing itself is loose, then they will fail the test. The rear lamps must be facing to the rear.

Examine the lenses to make sure they are not broken. In

Figure 2.1. Lens repair tape can be used to cover small cracks and breakages in lenses, provided it is the correct colour. Whether such a repair is acceptable is a matter for the tester's judgement.

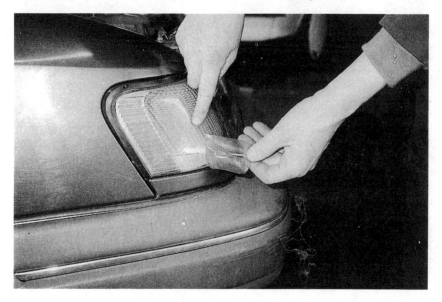

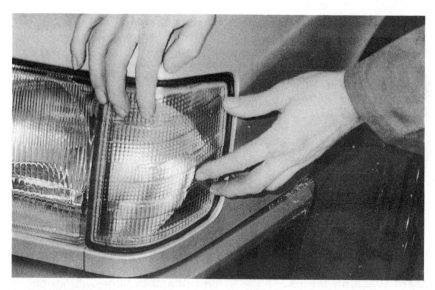

Figure 2.2. Lightly tap each sidelamp to make sure it does not flicker. If it does it indicates a poor connection or corrosion in the light unit.

particular, rear lamps must not show a white light to the rear. Repaired lamps with the proper repair tape are acceptable provided the tester is satisfied that the repair does the job properly (see Figure 2.1).

Colour and Brightness: The side lamps must show white to the front (or yellow if fitted into a headlamp which has a yellow light) and red to the rear. This is easy to check, but if any repairs have been made to broken lamps it is essential that the repaired lamps shows a colour that matches.

The size of both front and rear lights must be such that they can be seen from a reasonable distance. Any obscuring by other parts of the vehicle must leave at least 50 per cent of the lamp fully visible—both at the front and the rear.

It is easy to check for brightness and colour. In particular make sure that any repair tape is of the correct colour, and is not so thick that it dims the light output.

Accessories which have been fitted to the vehicle may obscure the lamps. Stand back from the car and see if there is at least 50 per cent of the lens showing to the front and rear.

Operation: The lamps must be operated by a properly working switch securely fitted to the car. There is no problem if the switch is different from that originally fitted to the car, as long as it is securely attached and works properly.

Each lamp must work consistently. After switching the lights on, the tester will lightly tap the lamp to see if it flickers (see Figure 2.2). If it does, then it will be failed. Flickering is usually caused by poor electrical connections, frequently resulting from corrosion in the light unit.

Dismantle the unit and check that all the electrical connections are good. If there is a problem, then bare the wires, and clean the ends before tightening up the connections.

A more common problem is an 'earthing' fault. This is when one or more of the lamps dim as other lamps—for example, the brake lights or the indicators—are operated. This results in not only the lamp being operated, but any other lamps affected, being failed. It is usually caused by corrosion affecting the electrical circuits. To solve the problem, remove and clean the assembly, ensuring that all metal to metal screws and connections which attach the lighting assembly to the vehicle are cleaned, and that all the wires make good connections, before putting it back together again.

The help of an assistant here is advisable (see Figure 2.3),

Figure 2.3. When checking the lights, an assistant is essential to point out any problems. This is particularly necessary when checking the stop lamps.

particularly when checking the rear lights, where earthing faults are commonly found.

Standing clear of the vehicle to see all of the lamps to the front or rear, in turn, ask the assistant to turn on the side lights, and then to operate the indicators in turn with the ignition switched on. When checking the rear lights the assistant should press the brakes on and off at the same time. Make sure that for each side, when the indicator is operating, the brake lights work correctly, *and* that none of the other lights is affected.

Finally, lightly tap each lamp in turn to make sure that it does not flicker. If it does, the problem is either a local earthing fault caused by the body of the lamp not being properly earthed, a loose wire in the local lamp circuit, or perhaps a loose bulb in a corroded or worn bulb holder. If all the lamps flicker, then the connections to the switch should be checked.

HEADLAMPS

You cannot accurately test the aim of your headlamp beams for the MOT without the special headlamp testing equipment used by the testing station (see Figure 2.4). Nevertheless, there are some aspects of the headlamp test which you can check, and even the aim can be roughly assessed to make sure that it doesn't deviate widely from the requirement, although this is more stringent now than in previous years, and your car may still be failed if the beam testing equipment shows it to be outside acceptable limits.

Both dipped and main beam, will have to be looked at, but it is not essential to have the main and dipped beam in the same lamps.

If the vehicle was used on or before 1 January 1931, then it does not require headlamps at all, but if they have been fitted they will be tested. However, the

Figure 2.4. The MOT garage will use a special piece of equipment to check the aim and beam pattern of the headlamps. Without this it is not possible to fully check your headlamps up to MOT standard, and the equipment is neither readily available to nor financially viable for the DIY mechanic.

testing requirement differs in that the dipping can be the 'old fashioned' style, where only one lamp needs to dip if the other switches off.

What the MOT test looks for
The main factors involved in testing the headlamps are:

* That the headlamps are a 'matched pair'.
* That the headlamps are

securely attached.
* Brightness, beam pattern and aim.
* Switch and operation.
* Lamp position.

How to check
A 'matched pair' of headlamps: Whether the headlamps are separate pairs for the main and dipped beams, or whether they are both contained in the same lamp assembly, they

Figure 2.5. A 'matched pair' of headlamps is essential. In practice this means they must be the same shape (side-to-side) and of the same level of intensity.

Figure 2.6. Check that the headlamp is not loose in its mounting, and that the housing is securely attached to the car.

must all be 'matched pairs' (see Figure 2.5).

There are two requirements for a 'matched pair' of headlamps. The headlamp beams must be of about the same intensity and colour and both lamps must be of the same shape and size, and be symmetrical to each other. In practice this means that the two sides must have the same type of lamp, whether or not main and dipped beams are in the same housing, or enclosed in different housings.

Make sure that the lights on both sides are the same shape and colour, and that the bulbs are of the same rating so that the intensity is the same from both lamps. Ironically, the way current regulations are written, it doesn't seem to matter if the *pattern* thrown by each lamp is not the same, and a mismatch of this kind would probably not be a reason for failure of the test.

The headlamp must be securely attached: Whether your vehicle has four headlamps—two for dipped beam and two for main—or just two—with the main and dip together—they must not be loose.

With the lights off, check that the headlamp sealed beam unit is not loose in its housing (see Figure 2.6). Put the flat of the hand over the lens and try to push the lamp unit in and out and up and down. It should not move at all unless it is the older sealed beam type of headlamp which is mounted on a spring-loaded bayonet type of fixing. Except for the expected springiness, if the lamp unit is loose it will be failed by the tester. On some cars the lamp unit is fitted into an outer housing that is in turn attached to the body of the car. If yours is like this, make sure that the housing is also nice and tight.

Some cars, notably the later Mercedes range, the Lada and many other makes have a facility to adjust the level of the beam from inside the car. On these cars there may be some up and down movement from springiness in the mechanism. This is not a problem.

Brightness: The official requirement here is that the lamps are not too dirty, and have not deteriorated or been damaged to the extent that they provide insufficient light to illuminate the road ahead, or that the beam image is adversely affected.

First make sure that the lens is clean and, if it is cracked, that the effect on the light it gives out, or on the shape of the image, is minimal. Unless the lens is so broken that it is virtually falling out, there should be no major change in light output and beam image. Also check that the bulb is not faulty.

Beam pattern: MOT testing stations have special equipment which enables them to inspect a vehicle's beam pattern and to measure the aim of the beams very accurately. You cannot duplicate this aspect of the test without that equipment, but some approximate checks can be made.

First inspect the reflector located behind the lens and surrounding the bulb. If it is so corroded that it changes the beam image, or reduces the light focused on the road ahead, then it will have to be changed (see Figure 2.7).

Park your car in front of a wall and switch the lamps on. Examine the intensity and shape of the beams to make sure that each is bright and of the same shape. Unfortunately it is not possible to be entirely certain that the shape of the beam image is correct without the proper headlamp testing equipment, but it will show if one side is different from the other, or the image is fuzzy, so that you can put things right before the test and avoid failure on this account

Figure 2.7. Examine the headlamp reflector carefully to make sure it is not so corroded that it seriously affects the shape of the beam or the intensity of the light emitted.

Figure 2.8. Although the DIY mechanic is unable to properly check the headlamps without the special equipment used at the MOT test garage, shining the headlamps against a wall will reveal any major discrepancies in intensity and aim.

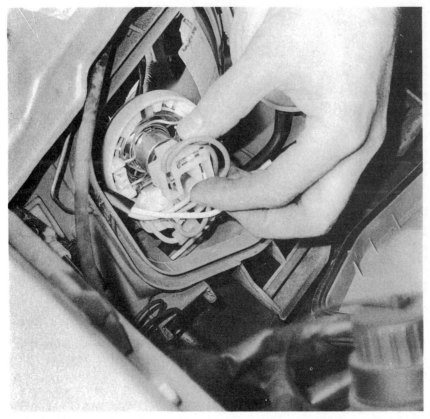

Figure 2.9. A fuzzy headlamp image can be caused by an incorrectly fitted bulb. Remove the bulb and make sure it is properly fitted with the locating lugs in place.

Figure 2.10. Check that the switch works properly and that the appropriate lamps operate in the correct switch position.

(see Figure 2.8).

If either the lens or the reflector need replacement on older cars, a complete sealed beam unit will have to be purchased as these components are not normally available individually. However, it is possible to buy the individual components (i.e. the case, glass, reflector and bulb) for more recent vehicles. If your headlamps do require attention, it is best to contact the local dealer for your make of car to see if the required components are available.

Another reason for a fuzzy image is an incorrectly fitted bulb. If the reflector is in good condition, and the lens is clean, with no cracks, but still the image is not sharp, try removing the bulb and refitting it securely into the special locating lugs (see Figure 2.9). This may solve the problem, but remember, it is not possible to accurately check the shape of the beam without the authorized equipment used by the testing station.

Headlamp aim: The regulations identify three different types of headlamp. Each has a different shape of beam, and each must be tested in a different way using the headlamp testing device.

This is something else you cannot accurately check at home. All you can do, without the authorized equipment, is to shine the beam against a wall, and make sure that on the main beam the beam is aimed along the car centre line, and that the two beams have the same shape of image on the wall. Then check that both lamps dip in the correct direction (towards the passenger side). Unfortunately that is all you can do. The headlamps may still fail on beam aim and shape, but you will have done your best to get them through.

Switch and operation: Both the switch and all the lamps must work on both dipped and main beam, and work correctly (see Figure 2.10). Here an assistant is useful, but not essential. If checking the lamps without assistance, you will need to aim them against a wall.

First switch on the headlamps and make sure that both sides come on at the same time, and that when the dip switch is operated, the beams do dip in the correct direction—to the passenger side of the road. (see Figure 2.11).

If one of the lamps fails to work, either on main or dipped beam, then a broken bulb is the most likely cause. Remove the bulb from the lamp and check it. Alternatively there could be a wiring fault.

Now lightly tap each lamp in turn with the beam switched on, to see if the light flickers. If it does, then it will cause your car to fail its test, so you will need to fix it. The problem will most probably be either a loose live wire or a fault somewhere in the car's wiring loom. The live connection is easily checked, but if the problem is in the wiring, then expert assistance may be required.

If the main switch does not operate properly, or has to be 'fiddled' with to turn the lights on and off, or to dip them, then it must be replaced (or repaired, if that is possible).

Lamp position: For production cars this will not be a problem. However, if your car has been fitted with auxiliary headlamps, they must be about the same distance side to side from the centre of the car, and at about the same height from the ground. It is not necessary to measure the distance—checking by eye is good enough.

STOP LAMPS

Stop lamps are not always required. If the vehicle was used before 1 January 1936, then it does not need stop lamps, but if

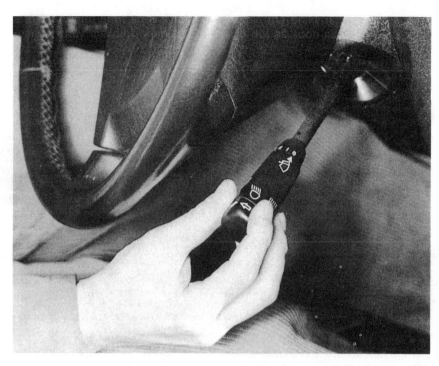

Figure 2.11. Operate the dip switch (in this instance on the indicator stalk) to make sure it works, and that it dips the headlamps properly. An assistant is useful for this check.

it has had them fitted they will have to be tested.

On vehicles used before 1 January 1971 only one stop lamp need be fitted to the driver's side of the centre line. Again if it has two stop lamps they will both be examined.

All other vehicles require two stop lamps. The examination of the stop lamps and reflectors is very similiar to that for the rear side lamps.

What the MOT test looks for
* **Security and condition.**
* **Operation and position.**

How to check
Security and condition: Both stop lamps (if fitted) must be properly attached and facing to the rear of the vehicle. They must not be incomplete, damaged, or have faded to the extent that they cannot be seen easily or do not work properly.

First examine the lenses of the lamps. If they are broken so that either a white light shows to the rear, or they are dim, then they

will be failed. However, repair with proper lens repair tape will be acceptable, provided the tester is happy with the result (see Figure 2.1 on page 13).

Now check the lamps for security. If they are loose in their housing, or the housing itself is loose, they will not pass (see Fig. 2.12). Nor will they pass if they are not properly facing to the rear, or more than 50 per cent of the lamp is covered by some other piece of the car (an auxiliary reversing lamp, for example).

Operation and position: The brake lights must illuminate immediately on application of the brakes. They must stay on while the brakes continue to be applied, and go out immediately the brakes are released.

An assistant will be needed for this part of the test. First make sure both lamps operate when the assistant applies the brake (with the ignition on!), stay on while the brake is held on, and go out as soon as it is released.

Figure 2.12. The stop lamp is normally an integral part of the rear light cluster. Check the assembly to make sure it is not loose and has no broken lenses.

Figure 2.13. The stop lamp bulb will be found in the rear lamp assembly on most modern cars. If only one side is operating, check the bulb and replace it if necessary.

If only one side works, there is a good chance that the bulb needs changing (see Figure 2.13).

On some cars the stop lamp switch is mechanically operated, and on others it is activated by a hydraulic pressure switch in the braking system. If both lamps fail to work, this switch is the most likely cause (see Figure 2.14). It will have to be replaced. The switch will also be the culprit if the lamps fail to go out straight away when the brakes are released. Again, replacement is normally the only remedy. If the switch is hydraulic, make sure that the braking system is properly bled after it has been replaced.

Now switch on the side lamps and operate the brakes to make sure that all the lights work together. As noted earlier, if the lamps flicker, or dim when the brakes are operated, or seem to be working incorrectly in any way, then an earthing fault is the probable cause. The indicators (flashing) must now be checked with the side lamps both on and off, and the brake lights being operated. Again, any problems not present when they were operated on their own indicates an earthing fault.

Finally, on this part of the examination, make sure that the stop lights are symmetrically located on the car both horizontally and vertically.

REAR REFLECTORS

All vehicles must have two reflectors fitted facing to the rear, and they must be red. They must be fitted more or less symmetrically on either side, be securely attached and be clean and undamaged such that they function properly.

What the MOT test looks for
The requirements here are:

* **Colour, security and prominence.**

How to check

Make sure that the reflectors are both red, and that the lenses are not so scratched or damaged that they will not properly reflect the light falling on them. Make sure they are firmly and tightly attached, and that they are not obscured by more than 50 per cent by any other parts of the vehicle when viewed from behind.

Any other extra reflectors are not part of the test, and reflectors formed by the use of reflective tape are not acceptable for the MOT.

DIRECTION INDICATORS

As for the brake lights, the direction indicator requirement depends on the age of the vehicle. If the vehicle was used before 1 January 1936, then indicators need not be fitted—but if they have been fitted they must be tested. Vehicles registered after 1 April 1986 must also have at least one side-repeater indicator on each side of the vehicle, and they must work, although those with a wrap-around lens which includes a side repeater are acceptable. These will have either an 'e' mark in a rectangle with the number '5' above, or an 'E' mark in a circle marked somewhere on the lens.

What the MOT test looks for

Surprisingly, the position and prominence of the indicators is not part of the test. What is checked is:

* Colour, security and condition.
* Operation.

How to check

Colour security and condition: The colour of the indicators must be amber if the vehicle was first used after 1 September 1965. Before that date the indicators can emit a white light to the front, and red to the rear. They must not be damaged so that the function is impaired,

Figure 2.14. On many cars the stop lamp switch is mechanically operated by the brake pedal as shown here.

and must be clean.

After determining the age of your vehicle, make sure that the indicators are of the correct colour, and that if side-repeater indicators are required, they are fitted. If a lens is cracked, it is assumed that a repair which will not impair the function of the indicator, or affect the colour of the light emitted will be acceptable, although the tester's manual offers no guidance on this.

In the same way as was described for the stop lamps and side lamps, check that they are not loose in the housing, and that the housing is not loose in the vehicle.

Operation: The indicators must all operate properly, and the switch must not be faulty in operation (see Figure 2.15). They

Figure 2.15. Operate the indicator switch while the assistant checks the operation of both front and rear indicator lights. At the same time check the 'tell tale' light.

must flash between 60 and 120 times a minute, and not be affected by the operation of any other lamps. If the operation of at least one of the indicators on either side of the vehicle cannot be seen by the driver so that he can tell that they are working, then some 'tell-tale' indication must be available to the driver to confirm that they are working. This can be audible or visual.

Ask someone to operate each side indicator with the ignition switched on. If the flash frequency is slow then start the engine and gently rev it up. The slow operation could be caused by low voltage, and the extra charge from the alternator or dynamo may provide more voltage to speed up the rate of flashing. This is allowed for in the MOT test. If the flashing rate is still too slow, then a new flasher unit will probably be needed.

As noted in the sections on stop lamps and side lamps, also operate the indicators with these lights on to check for earthing faults.

Make sure that the indicator switch works properly. If it is loose, or has to be 'fiddled' with to make the indicators work, then it will result in a fail. If an auxiliary switch has been fitted, it will be acceptable provided it works properly, and can easily be reached by the driver. Any self-cancelling mechanism is not part of the MOT test.

SEMAPHORE INDICATORS

Generally this will only apply to older cars where a small plastic illuminated arm extends from the bodywork on one side or the other to indicate which direction the driver intends to turn. The requirement here can be simply dealt with.

The switch requirement is the same as for the flashing indicators, but the semaphore must not stick, either in the

Figure 2.16. Very old cars may be fitted with a semaphore – 'trafficator' – instead of flashing indicators.

extended position, or fail to extend. It must illuminate when extended with an amber light to both front and rear. If the semaphore arms can be seen by the driver, then a 'tell-tale' is not required (see Figure 2.16). Otherwise there must be a 'tell-tale' which works.

HAZARD WARNING LIGHTS

Hazard warning lights are only required if your vehicle was first used on or after 1 April 1986. If they have been fitted to earlier vehicles, then they will have to be tested.

What the MOT test looks for
The test for the hazard warning lights is exactly the same as that for the flashing indicators, except:

* **The hazard lights must all flash in unison.**
* **There must be a 'tell-tale' that the driver can see inside the car.**

How to check
The hazard lights must all flash in unison: Switch on the hazard warning lights and walk around the vehicle checking that they are all operating properly, and flashing together at between 60 and 120 times per minute (see Figure 2.17).

The 'tell-tale' must operate correctly: Sitting in the driver's seat, make sure that the 'tell-tale' can be clearly seen and is working.

NUMBER PLATE LAMPS

The operation and security of these lamps will be checked in the same way as the side lights.

What the MOT test looks for
They must not be loose, they must work properly, and not flicker when tapped.

How to check
All vehicles must have at least one number plate light. If your car has more than one number plate light, then if any one bulb is not working, it will be failed (see Figure 2.18), so make sure *all* the bulbs are working.

REAR FOG LIGHTS

All vehicles first used on or after

Figure 2.17. The hazard lights must all flash in unison. Walk around the car to check that they work properly.

Figure 2.18. Make sure that the number plate lamps comply with the testing requirements. The most likely problem is a 'blown' bulb.

QUICK TIPS

* Check the bulb first, this is the most common cause of MOT failures on lighting.

* When changing bulbs make sure that they are properly fitted, and that the bulb holders are cleaned and any corrosion has been removed.

* On older cars replacement switches may no longer be available from the manufacturer. The next place to try is the local breaker's yard. If the proper switch is still not available, then a different 'universal' switch obtainable from a car accessory shop can be fitted. Remember that if a 'tell-tale' is required (for e.g., for the fog lamps), then a switch with an incorporated 'tell-tale' is the better alternative.

1 April 1980 must have at least one rear fog lamp fitted to the centre or offside (driver's side) of the vehicle. It must only emit a steady red light, and the 'tell-tale' inside the car indicating that it is switched on must be working. Check this light with the side lights and headlamps switched on.

Note: For all the lights, make sure that they can be properly operated together, where this applies, so that there are no interrelated faults or earthing problems.

Chapter 3

STRUCTURE AND BODYWORK

This chapter deals with the condition of the vehicle with regard to corrosion, damage and the security of the main structure and bodywork. The term 'bodywork' includes the doors, tailgates, dropsides (on commercial vehicles) and the seats.

It looks into the way the testing station examines the structure and bodywork of your car to decide if any corrosion, damage, distortion or insecurity is so bad that it will be failed. The criteria for failure is not always the same. A badly rusted part of the bodywork may be acceptable for the MOT test, whereas a similar level of corrosion in the main vehicle supporting structure, or close to a seat belt mounting, for example, would probably result in failure. For this reason the way the structure and bodywork are checked for the MOT test will need to be looked at separately. Consider the structure first.

STRUCTURE

It should be noted here that this is a difficult part of the test and the tester uses a lot of accumulated skill and knowledge to decide just which parts of the structure of any given vehicle need to be inspected, and how much damage or corrosion is acceptable before a failure results. The DIY mechanic is probably best advised to play safe and, if in doubt, repair it! Nevertheless, the specific requirements will be explained in some detail.

There are three aspects to the examination which have to be carried out on the structure for MOT test purposes. These are:

1. **The load bearing structure**. The main chassis or 'load bearing structure' of the vehicle (see Figure 3.1) will need to be examined to make sure that there is no corrosion, damage or distortion which in the opinion of the authorities would weaken the vehicle's structure sufficiently to make it unsafe to use on the road. The areas which can be regarded as part of the load bearing structure are quite specifically laid down by the Vehicle Inspectorate, and are referred to as '**prescribed areas**'. These will be looked at in detail shortly.

2. **Supporting parts of the structure**. Sometimes a part of the structure does not come into the definition of being 'load-bearing', but does contribute to the continuity of the overall strength of the

Figure 3.1. Shows the main front chassis member (A) which is a vital part of the main load-bearing structure. Clearly illustrated is the front lower suspension arm (B) which also carries the torsion bar spring (C). The arm is attached to the chassis member (D). This is an example of a 'prescribed area'. Any excessive corrosion would result in a failed test.

vehicle, and if it fails, steering or braking difficulties could result.

These are supporting parts of the structure, some crossmembers and strengthening beams would fall into this category (see Figure 3.1). In this case the testing rule is that excessive corrosion, damage or distortion outside the prescribed areas will result in a failure if it adversely affects steering or braking because it has severely reduced the strength or continuity of a structural member which has a main load bearing function. These areas are not specifically laid down, and the tester would use his judgement and experience to decide whether or not some supporting parts of the structure should be examined.

The best that you can do is to use common sense and make an assessment as to whether a crossmember or beam would adversely affect the steering or braking if it failed because of corrosion; or if it is distorted—so that this seriously affects the essential alignment of steering or braking components with adverse results.

3. **The structure within 30 cm of safety-related components**. A fail will also result if there is excessive corrosion, significant damage or distortion around certain parts of the vehicle which could be regarded as safety-related items. Steering and braking components are obvious candidates here, but this also includes suspension mountings, seat belt mountings and so on. A distance of 30 cm is set as the 'surrounding area' which has to be examined in the proximity of such components (see Figure 3.2).

Note: Because this part of the test of the structure will be carried out at the same time as the brakes, steering, seat belts and so on are being examined, it will be explained in the chapters dealing with those aspects of the test, so item 3 will not be further discussed in this chapter.

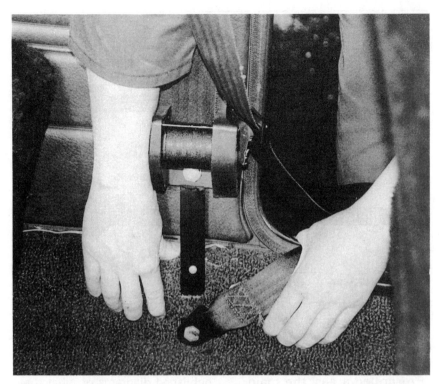

Figure 3.2. Some areas may be neither load-bearing nor supporting parts of the structure but still subject to test for corrosion, damage or distortion because they are within 30 cm of a safety-related component. This would apply to the area surrounding the seat belt mounting (above) and the telescopic shock absorber lower mounting (below), although here the criterion is whether or not the shock absorber is insecure or detached.

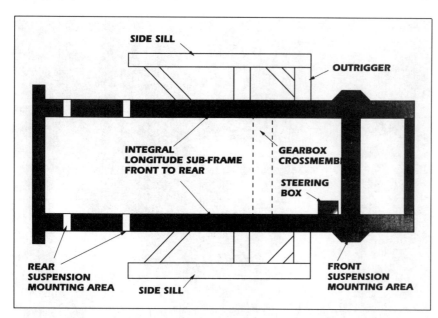

Figure 3.3. Structural integrity and corrosion: 'Prescribed areas'. (Reproduced with the kind permission of the Controller of Her Majesty's Stationery Office.)

'Prescribed areas' the main load-bearing structure

Different vehicles have markedly different structural designs, so there is no simple way of deciding how much of the structure is a 'prescribed area' and considered to be the main load-bearing structure of the vehicle. To address this problem the Vehicle Inspectorate have published diagrams of what they see as the more popular structural designs of vehicles subject to the MOT, and they have identified which areas fall into the testable category. These are shown in Figures 3.3, 3.4, 3.5 and 3.6.

Before even attempting to inspect your car for corrosion or damage of the structure, have a good look at the way it is constructed and decide which areas will need to be very carefully checked.

Unfortunately the diagrams do not provide a fully comprehensive guide, and for types of structure not included in the pictures the tester has to decide each case on its own merits. So, if in doubt as to whether a corroded or damaged part of the structure falls into the category of a 'prescribed area', repair it anyway to be sure.

Corrosion, damage and distortion

Corrosion: The technical expression used to indicate that a car has failed the MOT test because of corrosion is that it is 'corroded to excess'. This is important because some rust is inevitable on most cars over a few years old, but it is the extent of the corrosion (normally seen as rust) which is the crucial factor in deciding if any part of the vehicle will fail the MOT because it is 'corroded to excess'.

For structural corrosion the basic test is called the 'finger and thumb' test, because it is just that. The tester cannot apply more load to a suspect part of the structure than he can exert by squeezing the metal with his finger and thumb (see Figure 3.7). If the metal 'gives', or collapses, then it will fail the MOT. The only other check which can be made is to *gently* scrape or tap the suspect metal surface with a special light hammer approved by the testing authorities (see Figure 3.8). If holes appear in the surface then the metal is 'corroded to excess'. The tester is also allowed to tap the metal to see if the sound given off is dull, indicating the metal has been inadequately repaired with plastic filler (see Figure 3.9).

Damage and distortion: This refers to broken or distorted metal in the structure which

Figure 3.4. Structural integrity and corrosion: 'Prescribed areas'. (Reproduced with the kind permission of the Controller of Her Majesty's Stationery Office.)

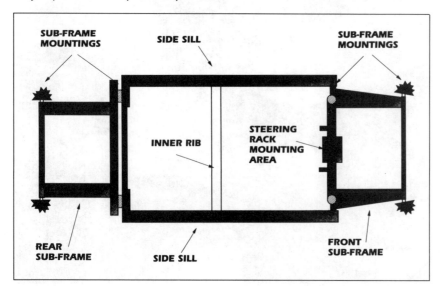

would weaken that part of the structure sufficiently to result in a fail. It could be accident damage, fatigue cracking, improperly carried out repairs and so on. There is no set method of inspection here, and normally it is fairly obvious when broken, cracked or badly distorted metal will result in a fail. Ultimately it is up to the tester to decide in each case. Sometimes the cracks are difficult to detect and the structure has to be loaded to highlight the problem. Very careful examination of the highly stressed parts of the structure will thus be needed to pick up these cracks.

What the MOT test looks for

Very old vehicles have a separate body and chassis, and in such cases the bulk of the main structure is to be found beneath the vehicle. However, with most modern cars that is not the case, and the structure of the car will be inspected at all stages of the test. First from inside the vehicle, then the underwing, rear boot, and underbonnet areas will be

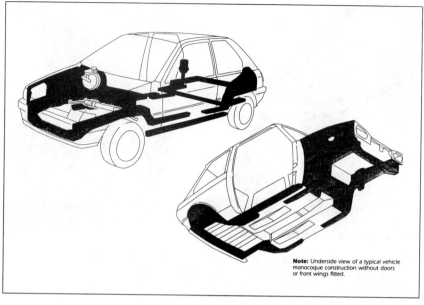

Note: Underside view of a typical vehicle monocoque construction without doors or front wings fitted.

Figure 3.5. Structural integrity and corrosion: 'Prescribed areas'. (Reproduced with the kind permission of the Motor Insurance Repair Research Centre.)
Note: Underside view of a typical vehicle monocoque construction without doors or front wings fitted.

Figure 3.6. Structural integrity and corrosion: 'Prescribed areas'. (Reproduced with the kind permission of the Motor Insurance Repair Research Centre.)
Note: Underside view of a typical vehicle monocoque construction without doors or front wings fitted.

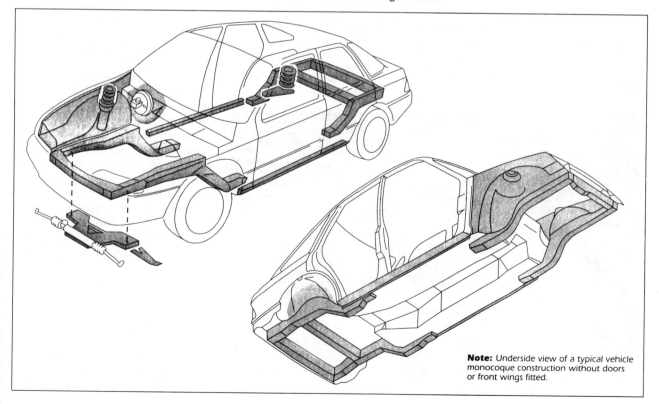

Note: Underside view of a typical vehicle monocoque construction without doors or front wings fitted.

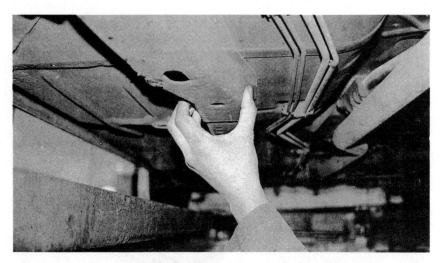

Figure 3.7. The only load which can be applied by the tester to check the strength of metal which is corroded is that produced by squeezing the suspect area between fingers and thumb.

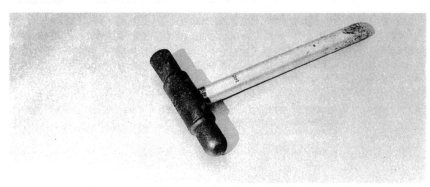

Figure 3.8. The special light hammer/scraper approved by the test authorities for use in detecting excessive corrosion.

Figure 3.9. Gently tapping with a light hammer can help to determine whether structural damage has been adequately repaired with plastic filler. The filled areas will sound 'dull' when tapped.

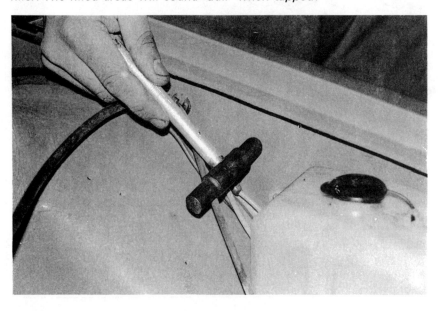

examined. Finally the whole structure beneath the vechicle will be thoroughly checked.

Taking these in turn:

Inside the car
The main areas checked here are the inner sills which are just below the doors inside the car, and run the length of the car. It should be emphasized that these will only need to be tested if there are no other structural beams which run along the whole length of the car in the same relative position along the vehicle. Generally the floor will also need to be checked, but that is not necessarily the case. So, if applicable, what needs to be tested are:

* **Inner sills.**
* **Floor.**
* **Suspension component mountings (for vans and estate cars where applicable).**

How to check
Inner sills: Grasp the sills in your hand and squeeze to see if the metal has any give (see Figure 3.10). Also knock the sills all along their length, this may indicate that filler has been used. It must be noted here that unless the carpet can be easily removed, the tester is not allowed to lift it to carry out this test, but you will have no such restraint, and removing the carpet will soon reveal the extent of corrosion.

On many vehicles the seat belt mountings are attached to the inner sills, and it should be remembered that part of the test here will be to see if the structure within 30 cm of the seat belt mounting is excessively corroded.

If the sill is bent or slightly dented, but the metal is still solid, it is not likely to be failed.

Floors: Normally the floor can be inspected from beneath the vehicle, and with no carpet in the way it is easier to inspect, but even so it is worth having a good check around the floor areas inside the car to see if there is any excessive corrosion.

Suspension component mountings: Inside the car this only applies to estate cars and vans where there is not a boot. On some vehicles the upper rear shock absorber mountings penetrate into the vehicle itself. This area must be examined for corrosion which has caused the shock absorber to become detached or insecure. Push firmly all around the suspension mounting (see Figure 3.11).

Also, in these areas two or more layers of metal may be found, and although the metal feels strong and passes the 'finger and thumb' test, any corrosion between the layers may have resulted in weakness. The effect of this type of corrosion is to make the upper layer of metal lift, and this can be seen quite readily. Whether a failure will result or not is a matter of judgement as to whether or not the shock absorber is insecure, but if in doubt, repair it.

Under the bonnet and inside the boot

From the diagrams of different vehicle structures shown in Figures 3.3 to 3.6 it is clear that on some types of structure almost the whole under-bonnet area will need to be checked, whereas on others only specific parts will need to be examined. Once again, the same advice applies—if in doubt, repair it. The areas checked here are:

* **Main load bearing parts of the structure, the 'prescribed areas'.**
* **Suspension component mountings.**

The 'prescribed areas': Having decided which areas on your vehicle could be called 'prescribed areas', using your hands only, push and pull at any

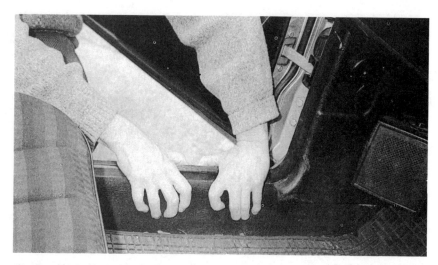

Figure 3.10. If your vehicle does not have structural beams which run the length of the car, then the sills form part of the main load-bearing structure and have to be tested. Try to squeeze the metal. If it gives or can be heard to 'crunch' then it is probably corroded to excess and would fail the MOT.

obviously corroded areas. Have a good look for any damage and distortion. If necessary you can give any suspect patches of corrosion a quite hard tap with a small hammer, or a good poke with a screwdriver since you are not constrained in the way that the tester is.

Always look very hard at places where two pieces of metal have been overlapped, corrosion often starts in the crevices which

can trap water between them. Metal around the battery tray is also frequently corroded where there has been spillage at some stage (see Figure 3.12), although this is not a failure unless it is in a 'prescribed area'.

Suspension component mountings: This test is the same as that described in the last section where the inspection was carried out from inside the car. A

Figure 3.11. Inside estate cars and vans the upper shock absorber mountings can be seen and inspected for corrosion which has caused insecurity or complete detachment of the shock absorber.

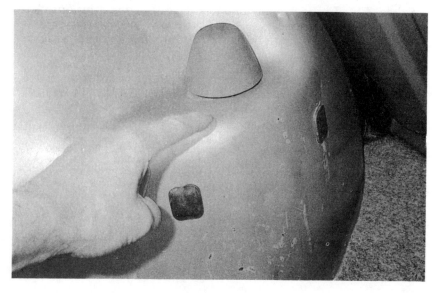

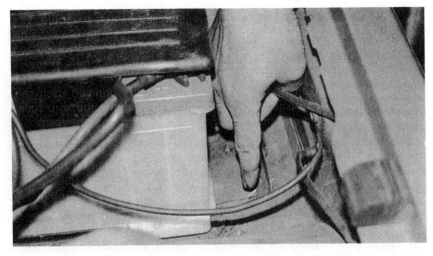

Figure 3.12. Check particularly carefully around the battery tray since this can be prone to corrosion.

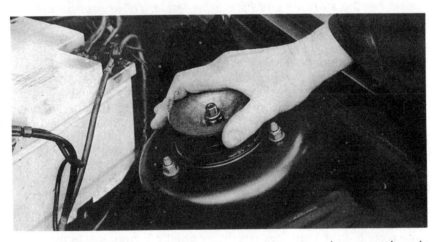

Figure 3.13. The metal around under-bonnet suspension mountings is frequently found to be corroded through water, splashed up from beneath the wing, penetrating the metal layers.

Figure 3.14. To examine the under-vehicle structure it may be necessary to put your car on ramps (as shown here), jacks or jack stands. Always chock the wheels on the ground, and use safety jack stands.

lot of cars have the main upper front suspension mounting located beneath the bonnet, and this can be subject to corrosion and delamination of the doubled metal (see Figure 3.13). Normally any problems can be easily spotted by a simple visual examination.

Under-vehicle inspection

The MOT testing garage will have a vehicle inspection lift or pit to carry out this part of the test, but you will probably have to make do with a jack and jack stands, or a small ramp. While underneath the car it is essential to take care, and ensure that the car is safely on jack stands or located securely and safely on the ramps.

ESSENTIAL SAFETY PRECAUTIONS

For safe under-vehicle inspection make sure that the handbrake is firmly applied, that the wheels on the ground are securely chocked, and that any ramps, jacks and jack stands are firmly located on flat and level ground before lying beneath the car (see Figure 3.14).

Safety inspection lamps: It will be necessary to have some form of inspection lamp for this part of the test. There are many such lamps available which operate off the car's cigarette lighter socket and these are very good and perfectly safe.

Do not use a mains lead lamp: This is because any earthing fault, or water around the car could result in electrocution.

How to check

The under-vehicle inspection can be broken down into three main areas, and these are:

* **The front structure, including the main**

load-bearing structure carrying the front suspension.
* **The main structure supporting the vehicle along its length.**
* **The rear structure, including the main load-bearing structure carrying the rear suspension.**

Before carrying out this part of the examination of the car for the MOT, have a good look at the diagrams of different types of structure shown in Figures 3.3 to 3.6, and decide if your car falls into one of those categories. Otherwise, have a good look at your car's structure and see which parts are obviously designed to take the main loads.

Front structure: From beneath the car have a look around the front. There will invariably be some suspension components which are connected to the main structure. There could be anti-roll bar mountings (see Figure 3.15), suspension ties, or perhaps the front spring mountings if leaf springs are used. If these are connected to the structure, or a front cross member, then that will have to be inspected to make sure it is not excessively corroded, damaged or distorted. This will apply to the whole of the structure concerned.

On some vehicles, notably the Minis, the front suspension is fitted to a sub-frame which is mounted onto the main structure. Not only will the mounting areas need to be checked, but also the condition of the sub-frame itself is part of the structure which will need to be examined for MOT purposes.

Moving towards the front wheels from the front of the car, the main structure will almost certainly carry the various suspension components, the inner suspension arm mountings, the steering box or steering rack mountings (see Figure 3.16) and

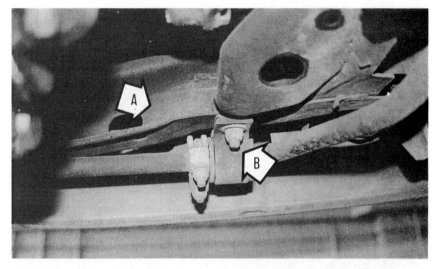

Figure 3.15. The front chassis crossmember (A) will need to be examined for excessive corrosion. This is specifically the case if it carries an anti-roll bar mounting, as shown here (B).

perhaps the shock absorber mountings. Check particularly carefully around this area where the steering and suspension is mounted to the vehicle. There will almost certainly be an accumulation of mud or dirt. Scrape this away in suspect areas before closely examining the

metal.

The main structure: Working back towards the rear, check the main structure. On some vehicles this will be in the form of chassis beams or sections attached to the main body shell. In other cases the whole of the floor and

Figure 3.16. The main structural crossmember shown here carries both the steering rack and the inner lower suspension joints. It will need to be thoroughly checked for corrosion.

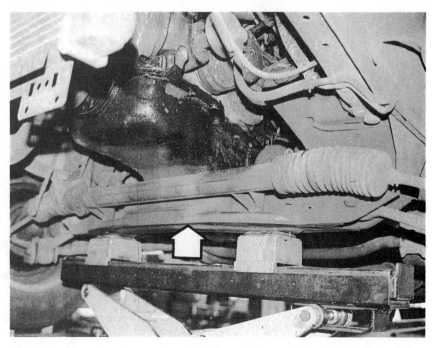

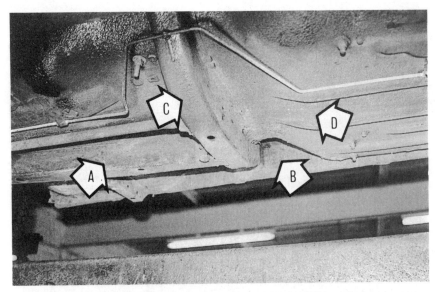

Figure 3.17. Where the main chassis beam runs beneath the car (A) the outer and inside sill (B) are not part of the main load-bearing structure. In the above case the main chassis stops at the main chassis crossmember (C). The other side of the crossmember, the floor, outer sill (not seen here) and inside sill form the main load-bearing structure (D), and they need to be examined for excessive corrosion, damage and distortion.

Figure 3.18. This shows a corroded outer sill (A), but because there is a main chassis beam (B) running behind the corroded part it would not be subject to test. The chassis beam stops at the crossmember (C) and the sill would be subject to test rearwards of that point.

structure beneath the doors will comprise the main structure, and every part will need to be examined for damage, distortion or corrosion (see Figures 3.17 and 3.18).

If any part is obviously corroded, in the form of rust, have a good look at it, and after scraping away the surface rust, try to squeeze the affected areas to see if there is solid metal beneath. Any sign of 'give' or rust holes will result in a failure of the test (see Figure 3.19).

The rear structure: The examination here is much the same as for the front. Find out where the suspension (including springs) and shock absorber mountings are located and check these areas carefully (see Figure 3.20). It should be noted that on some cars with independent rear suspension, there will be a complete subframe carrying the suspension. This is also the case on the Mini series (as at the front), and this will be attached to the main load-bearing part of the structure. Very carefully

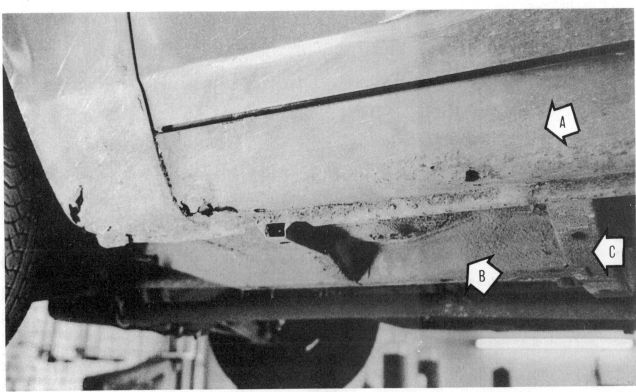

Figure 3.19. Corrosion on this scale, where a rust hole has developed, would result in a fail in any parts of the structure subject to test.

examine the areas around the attachment points, as well as the sub-frame itself (see Figure 3.21).

Sometimes the main structure extends to the very rear of the vehicle carrying the rear spring hangers when the vehicle is fitted with leaf springs. This area is very prone to corrosion because of the accumulation of water, road dirt and grime (see Figure 3.22).

Figure 3.20. When examining the rear structure, check very carefully around the rear suspension mountings as shown here. (A) Rear main chassis member. (B) Front mounting of rear leaf-spring. Note: Because there is no main chassis member running along the car ahead of the spring mounting, the condition of the rear floor area (C) would need to be examined for the test.

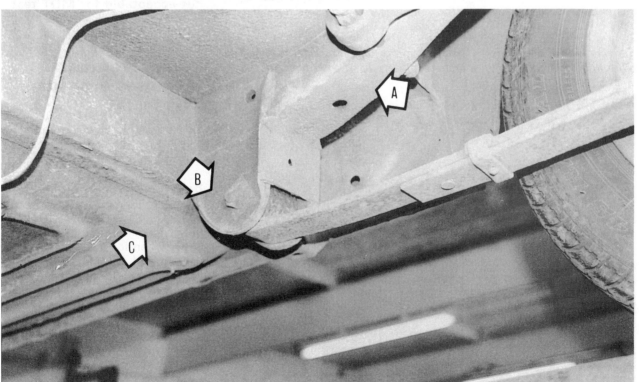

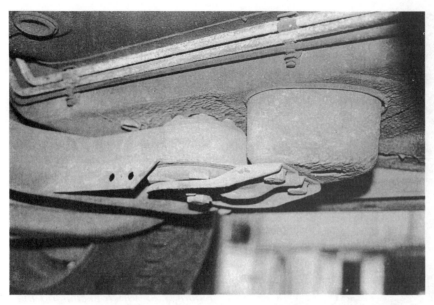

Figure 3.21. On some cars the suspension is not mounted directly to the chassis, but to a rear sub-frame which in turn is attached to the vehicle structure. Check for corrosion around the mountings and on the sub-frame itself.

On vehicles with complex rear suspension systems, there could be many connecting arms from the stub axle carrying the wheel, connecting back to the main structure. These are used to stabilize the rear suspension. If this is the case, then any beams or components of the structure to which such components are attached will need to be fully examined.

Welded structural repairs
If the structure is damaged or corroded to excess, then the repair will have to be properly welded to pass the MOT test. Repairs which have metal which has been riveted or bolted in place is not acceptable, nor is the use of fibreglass or other form of plastic filler. Sometimes smaller rust holes can be repaired with a metal patch welded into place. If that is done, then the metal used must be thick enough to effect an adequate repair, and the patch must be seam welded into position.

With other repairs—door sills for example—if the original component was welded into place with spot welds, then the replacement part can also be spot welded. But be careful, sometimes the replacement sill does not attach along the same seam as the original. In that case it will have to be seam welded into place (see Figure 3.23) and if it is thick enough to do the job, then that will be adequate for MOT purposes.

Note: The welded repairs of highly stressed items, such as suspension trays and arms, is not acceptable for MOT test purposes.

BODYWORK (Including doors, tailgates, dropsides and seats)

This aspect of the test involves checking for corrosion and damage, and also for security.

Bodywork (corrosion and damage)
This requirement of the MOT test only came into force in January 1993, when for the first time the condition of the bodywork had to be inspected. The criteria for part of the bodywork failing the MOT test is quite different from that for the structure and is very simple and uncomplicated. The tester simply has to assess whether or not the bodywork of the vehicle is so damaged or

Figure 3.22. Check the whole length of the main chassis structure to the very rear of the vehicle. The rear spring hangers at the leaf-spring mountings shown here are attached to the rear of the chassis. This area is particularly prone to rust.

Figure 3.23. A patch repair has to be seam welded, as shown on this door sill.

corroded that a sharp edge or projection is dangerous to other road users. This includes pedestrians. What the tester is looking for here is:

* **Accident damage which has caused a dangerous projection or sharp edge which could cause injury.**
* **Corrosion which leaves**

sharp or jagged edges which could cause injury.

How to check
Accident damage: It is only possible to provide general guidelines in this section, and the tester has to decide in each individual case. However, some situations can be visualized. If one of the body panels has been

Figure 3.24. Although the bodywork has been forced inwards here, the car may be failed because the end of the bumper has been left projecting out.

crumpled leaving a sharp projection at a level where it could cause injury, then your car will be failed (see Figure 3.24). Perhaps one of the most common causes of failure here is when the bumper end caps are missing, leaving sharp projections (see Figures 3.25 and 3.26). These projections could catch a pedestrian or cyclist, causing serious injury.

Corrosion: With corrosion the main issue is how extensive it is, the form it takes and where it is located. For example, the lower section of door panels sometimes corrode so that they detach from the door and leave a sharp knife-like protruding edge—this would clearly cause your car to fail its test (see Fig. 3.27). On the other hand, a large rust patch on the top of the front wing (also quite common) would not result in a fail because it is not likely to cause injury (see Figure 3.28).

Bodywork (attachment to main structure)
This only applies to vehicles which have a separate body and chassis (main structure). This would be the case in very old vehicles, and some light commercial vehicles. The regulations require that the body must be properly secured to the vehicle structure. In practice this means that:

* **All the mountings will need to be examined, both for security (are the nuts and bolts tight?) and for corrosion or damage.**
* **Is there relative movement between the body and the chassis?**

How to check
Mountings: Although the mountings are part of the test, the regulations are lenient in this area. If there are a number of attachment points securing the body to the chassis, then they do

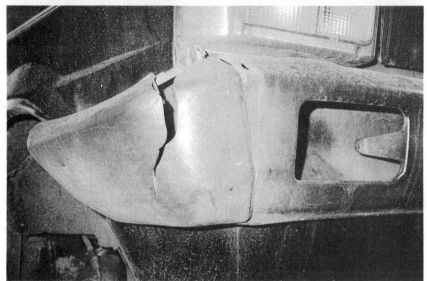

Figure 3.25. **Left.** *With the bumper end-caps missing, as in this case, the car will be failed because the sharp projections could cause injury.*

Figure 3.26. **Below left.** *Although this bumper end-cap has not become detached, it is for the tester to decide if it would be likely to cause injury. In such a case the owner would be best advised to play safe and have it replaced before the test.*

Figure 3.27. **Bottom left.** *If a rust hole with a sharp projection likely to cause injury is covered with sturdy tape to render it safe, then it will be accepted by the tester.*

Figure 3.28. **Below.** *Even serious corrosion on the top of the wing, where it is not likely to cause injury, would not cause a car to be failed.*

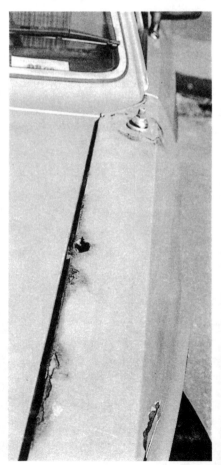

not *all* need to be secure with all the attachment nuts and bolts intact and tight, or *all* free from corrosion or damage. A few attachment points which are unsatisfactory when taken alone are acceptable, provided the cumulative effect will not be likely to significantly reduce the body/chassis security in overall terms.

Relative movement: Ask someone to push and pull the body of your vehicle while you examine the relative position of chassis and body structures. This will normally be possible without crawling beneath the vehicle. Any movement will be clearly evident and will result in a failure of the test if not repaired. Generally it is unlikely that a vehicle would pass the test on one aspect of this section, and not on the other.

Doors, boot lids, tailgates and drop sides
This aspect of the MOT test can be dealt with very simply and directly.

Doors: Both the driver's door, and the front passenger's door must close properly and stay closed on the latch. They must also be openable from both inside and outside the vehicle (see Figure 3.29). All other passenger doors need only latch in the closed position.

Boot lids, tailgates and dropsides: All these must securely latch in the closed position. If a light commercial truck is fitted with a detachable tailgate or drop side, and that has been removed, then it will not fail.

Securing such items with rope, webbing and the like is unsatisfactory for the MOT test.

Note: If a commercial vehicle is presented for test with an insecure load, then the tester can refuse to carry out the test.

Figure 3.29. Both the driver's door and passenger front door must be openable from inside and outside the vehicle. All other passenger doors need only latch in the closed position.

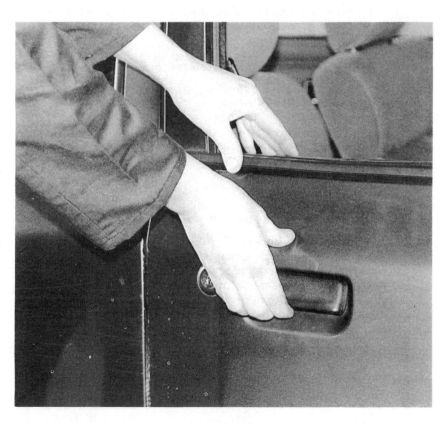

Figure 3.30. Both front seats have to be checked to make sure their security of attachment is not 'significantly' reduced.

Seats

As part of the vehicle structure, the seats are included in the MOT test. The requirement is that both the driver's and passenger's front seats have to be tested for security (see Figure 3.30). On the basis that most seats will move if vigorously pushed and pulled, the test only needs the seats to be checked to make sure that their security is not *significantly* reduced. This check also requires that the seat back is also examined to ensure that it can be secured in the 'normal' upright position (whatever 'normal' means!)

QUICK TIPS

The load-bearing structure
Find out which parts of the structure need to be checked:

1. **Check the diagrams of vehicle structures and check which part of the structure could be considered 'load-bearing' on your car.**

2. **Have a look at cross members, strengthening plates and so on, and ask yourself: 'Are they supporting parts of the structure providing continuity to the strength of main load-bearing areas?'**

Areas within 30 cm of a safety related component

1. **Does the metal being examined provide supporting strength within 30 cm of:**

* **a steering/suspension component;**

* **a braking component;**

* **seat belt anchorages.**

Welded repairs
1. **All patch repairs must be seam welded.**

2. **If the repair is welded along the same line as the original was spot welded, then spot or 'stitch' welding is acceptable. *Note:* As a rule of thumb, to be absolutely safe, use stitch welding in these areas making sure that 50 per cent of the length is actual weld.**

   If in doubt, then seam weld.

SAFETY CAUTION: Check the safety notes on Pages 6–8.

Chapter 4

STEERING AND SUSPENSION

There are many different types of steering and suspension systems fitted to motor vehicles, and it would take too long to explain in detail the MOT testing procedure for each individual system, but most of the common types will be referred to in the text and illustrated. The main point to remember is that whatever type of system is fitted to your car, *all* the joints and bushes of both the steering and suspension will be examined and must not be 'worn to excess', otherwise your car will fail the MOT.

On all modern vehicles fitted with independent front suspension the steering and suspension are interlinked and on some systems the swivel joints and the suspension joints are one

and the same. Although, in practice, the tester will examine the steering and suspension at the same time, to make it easier to describe they will be looked at separately here.

STEERING

Before explaining how to test the steering it is worth briefly explaining certain aspects of steering systems to be found in different designs of cars.

Steering mechanisms
There are two different types of steering mechanisms fitted to vehicles subject to the MOT test. These are the steering box and the steering rack. The main difference between the two is in

the way the movement of the steering wheel is transferred to become movement of the steered front wheels. Let's have a look at each in turn.

The steering box: In this system the steering wheel turns a shaft which is connected to a small gearbox called the 'steering box' which is attached to the vehicle structure. The shaft turns a worm gear inside the steering box which meshes with another gear to turn a lever one way or the other as the steering wheel is turned. The lever is called the 'steering drop arm' (see Figure 4.1).

This arm is then connected by a jointed rod (called a 'track rod') to an 'idler arm' assembly located on the other side of the car opposite the position of the steering box which is in effect another 'steering drop arm' but without the steering shaft and gears (see Figure 4.2).

As the steering is operated the idler arm mimics the movement of the steering drop arm. This side to side movement is then transferred to the wheels either side of the car by rods from both the steering drop arm and the idler arm. In turn these are connected to the components carrying the front wheels (called the stub axles) enabling them to swivel. These rods are described as tie rods, and the joints at their outer end which connect to the stub axle are commonly known as 'tie rod ends'.

Note: On some light commercial vehicles the drop arm movement is transferred through another linkage before it is connected to the track rod.

The steering rack: The steering rack is a single mechanical device which substitutes for the steering box, the track rod and the idler arm (see Figure 4.3). It is more correctly known as a rack and pinion steering mechanism. The steering shaft is

All cars subjected to the MOT test will have wear in their mechanical components. Whether or not a component is 'worn to excess'—the expression used in the test to indicate a failure—will depend on the extent of wear acceptable in *that* component on *that* car, and the subjective opinion of the tester.

The 'Inspection Manual' notes that the wear or 'play' is excessive if replacement, repair or adjustment of the component is necessary. Generally this is a matter of common sense and will be obvious to the DIY mechanic, but it is impossible to give more than approximate guidance here, so do not be too surprised if the tester fails an item you thought would be acceptable, or passes some joints which may seem to have had excessive wear but are in fact acceptable.

Figure 4.1. The steering box. In this mechanism the steering shaft inside the steering column (A) transfers movement from the steering wheel through gears located inside the steering box housing (B) into side to side movement at the steering drop arm (C). This is transferred to movement at the front wheels through various steering linkages. The steering box housing is connected to the chassis. Adjusting nut (D).

connected to a small toothed gear wheel (the 'pinion'), the teeth of which engage in similar teeth in a straight piece of metal called the 'rack' which can slide from side to side. This whole mechanism is housed in a long cylindrical shaped body which is connected across the car's chassis. This is positioned so that the ends of the rack are equally spaced across the car, with the pinion obviously located on the driver's side where it is connected to the steering shaft.

As the steering wheel is turned, the steering shaft turns the pinion which causes the rack to move from side-to-side. The tie rods are connected to the ends of the rack by swivel ball-joints concealed inside rubber concertina shaped 'steering rack gaiters' on each side, and these are then connected to the stub axles with tie rod ends in the same way as for the steering box mechanism (see Figure 4.4).

Steering swivels

The front wheels can be made to swivel in various ways to enable them to steer, and this can vary from car to car. The important point here is that all types of steering swivel systems can use either a steering box or a rack and pinion to transfer the

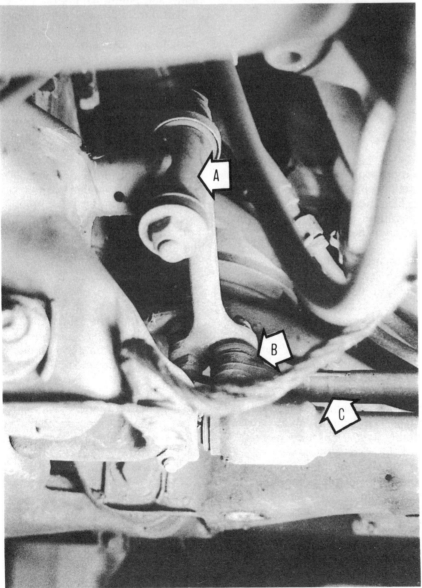

Figure 4.2. Idler arm assembly (A). All joints and pivots must be checked for excessive wear. (B) Tie rod end. (C) Centre rod.

Figure 4.3. Steering rack. Movement at the steering wheel is transferred by the steering shaft through a coupling to the splined spigot (A) which then rotates with the steering wheel. This is connected to a small gear-wheel inside the rack housing which meshes with teeth along the steering rack, resulting in side to side movement. This is transferred to movement at the wheels through tie rods (B). The steering rack is mounted onto the chassis. Steering rack rubber gaiters (C).

steering wheel movement via the track rods to the wheels.

It is also worth noting at this stage that on a lot of cars the front suspension and the steering swivels are one and the same, whereas on others the suspension joints are separate from the steering swivel joints. When examining the steering and suspension it will be vital to have a good look at the system to check this (see Figures 4.5 and 4.6).

Steering joints and connections
At each connection of the steering rods, drop arms, swivels, etc. there will be some form of joint, this could be a metal

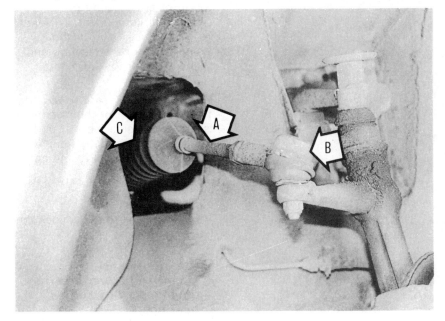

Figure 4.4. The side to side movement of the steering rack is transferred to the wheels through the outer tie rods (A) and tie rod ends (B). This picture also shows the steering rack gaiter (C) concealing the inner tie rod end. (The car is a Morris Marina.)

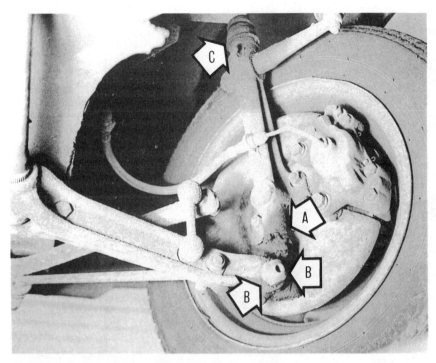

Figure 4.5. On this Morris Marina (Ital) the lower steering swivel joint (A) is distinct from the lower suspension joint (B), whereas the upper steering swivel and suspension joints (C) are one and the same. Any combination may be found and it is vital to know how the steering/suspension system works to properly examine all the joints for the MOT test.

Figure 4.6. On this car the lower suspension joint and the lower swivel joint are one and the same in the form of a ball and socket.

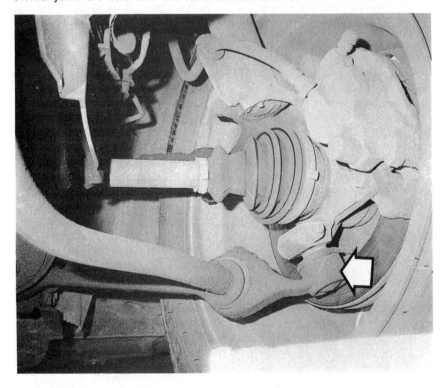

'ball-joint', or sometimes a special rubber bush. All these will have to be fully examined when your car is MOT tested.

Power steering

If your car is fitted with power steering, then this is a testable item and will have to be fully examined as part of the MOT test of the steering. Power steering systems are operated by hydraulic fluid under pressure providing power assistance as the steering is turned. All the pipes, hoses and hydraulic unions will have to be checked for the MOT test. Although the fluid level in the reservoir is not part of the test in itself, it is worth checking since if it is low a leak in the system could be indicated which would result in a failure.

It will also be necessary to ensure that the power assistance is working uniformly as the steering wheel is turned from lock to lock. The examination of a power assisted steering mechanism must take place with the engine running.

Corrosion and structure

As discussed in Chapter 3, the MOT test for corrosion or structural damage requires that this must not be excessive within 30 cm of any steering or suspension component. This will be an important part of the examination of the steering and suspension for the MOT test.

What the MOT test looks for

The examination of the steering takes place inside the car, under the bonnet, and finally from underneath the front of the car. Taking these in turn:

Inside the car

This part of the test examines the following items:

* Steering wheel.
* Upper and lower steering column bushes and/or bearings.

* Steering column
  attachment brackets.
* Steering couplings.
* 'Free play' in the steering.
* Lock to lock freedom of
  movement.

How to check

The steering wheel: Make sure
the steering wheel is sound by
pushing and pulling it to check
that it is not broken. Any breaks
or cracks in the wheel which
may 'catch' on the driver's hand
will cause it to fail the test. It
will also fail if a spoke is broken
and it is difficult to turn the
wheel without it collapsing.

**Steering column bushes and
attachment brackets:** Common
causes of failure are the steering
column bushes. The upper
bushes can be checked by
grasping the steering wheel and
trying to push it up and down at
right angles to the line of the
steering column (see Figure 4.7).
Excessive wear is identified by
obvious movement of the wheel
in relation to the rest of the car.
The lower bushes are not so easy
to check, but on most modern
cars which do not have a
steering column outer casing,
grasp the shaft as close to the
bush as possible and see if it has
movement indicating that the
lower bushes are worn.

Make sure that any movement
is in the bushes and not caused
by loose attachment brackets.
These can be checked in much
the same way, that is by looking
for movement when the shaft is
grasped and pushed and pulled
in all directions. If the brackets
are loose, corroded or even
broken they will fail the test.

**Steering couplings and joints
inside the car:** Any couplings
or joints in the column which are
located inside the car will also be
checked at this stage (see Figure
4.8). This is done by examining
the joint or coupling and making
sure that there is not excessive
wear or, if the joint is splined

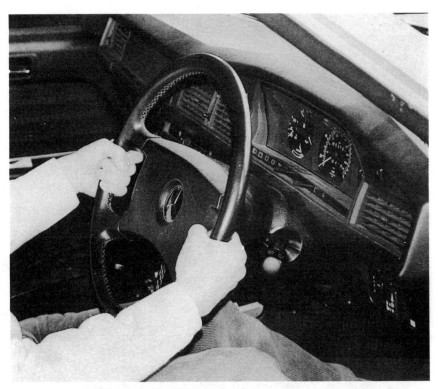

*Figure 4.7. Movement of the steering wheel relative to the steering
column could indicate excessively worn upper steering column
bushes, or the upper attachment bracket. Pushing and pulling the
steering wheel in all directions at right angles to the line of the
column will reveal any problems. Further investigation may be
required to determine if worn bushes or a loose mounting bracket is
the cause.*

Figure 4.8. This picture, taken beneath the dash panel of a Morris Marina, shows the items which will need to be examined in connection with the steering column (A) and steering shaft (B). The mounting (C) will need to be checked, as will the flexible rubber joint (D). Many cars will have two flexible joints in the steering shaft – one beneath the dash panel and another under the bonnet. Both must be thoroughly checked.

onto the steering column, that the splines are not worn to the extent that there is radial 'play'. With splined joints there will probably be a pinch bolt; make sure it is properly tightened (see Figure 4.9).

Figure 4.9. Where the steering shaft is connected to steering couplings, or the steering rack, make sure any pinch bolt which secures the joint (arrowed) is in place and tight.

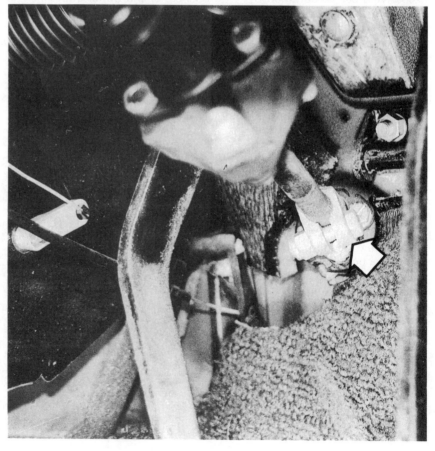

Note: It is a rule of the MOT test that no dismantling of components is allowed. So, if part of the steering mechanism is hidden behind a panel it is not examined as part of the test. However, if the wear is detected by (for example) movement at the steering wheel, but the worn testable item is concealed, then it will still fail.

'Free Play' in the steering mechanism: With the steering wheel 'centered' so that the front wheels point ahead, first turn the steering wheel one way until the front steered wheels just move, and then the other way also until the front wheels just move (see Figure 4.10). If there is excessive movement at the rim of the steering wheel before the front wheels respond, it will fail the test.

The amount of allowable movement depends on the type of steering mechanism fitted. If the car has rack and pinion steering, movement of up to 13 mm is allowed on a conventional 380 mm diameter wheel. For cars with a steering box fitted, the movement can be up to 75 mm. If the steering wheel is smaller or larger than normal, these limits vary accordingly.

If there *is* too much movement, then excess wear in some part of the steering mechanism is indicated and the cause will become apparent on further examination.

Lock to lock movement: During an MOT test the tester will turn the steering wheel from lock to lock to make sure that the mechanism works smoothly without 'notchiness', and that the resistance to movement is the same in both directions (particularly on cars with power steering). It is also important to check that the wheels do not foul any parts of the bodywork or structure on full lock either side.

At the testing station this is

Figure 4.10. Check for 'free play' in the steering by turning the steering wheel one way and then the other until the front wheels just move. Measure the movement at the steering wheel. More than 13 mm will fail if the car has rack and pinion steering. For a steering box up to 75 mm is allowable.

done with the front wheels placed on low friction turning plates (see Figure 4.11). Unfortunately you won't be able to fully duplicate this aspect of the test without those turning plates. Even so, it is worth asking

someone to turn the steering wheel gradually on to full lock one way and the other while you check the wheels, although it may not be possible to achieve full lock because of the friction of the tyres on the ground

Figure 4.11. At the MOT station the wheels are turned from lock to lock using special turning plates. This detects any 'notchiness' in the steering mechanism and also shows up where the wheels may foul on parts of the vehicle structure.

preventing the wheels turning those last few degrees.

Under the bonnet

The extent to which the steering system can be examined from under the bonnet will depend on the design of the vehicle being tested. On some vehicles it is impossible to see any of the steering components from beneath the bonnet, whereas on others the lower steering shaft coupling may be visible, or perhaps the steering rack gaiters. So have a good look around beneath the bonnet and examine any parts of the steering mechanism which can be seen. The most likely components which will be visible are:

* **Upper swivel joints and mountings on vehicles with 'MacPherson Strut' suspension.**
* **Steering rack rubber gaiters.**
* **The coupling from the steering shaft to the steering rack.**

* **The steering box and idler arm bearings.**
* **Corrosion and structural damage.**
* **Power steering.**

How to check

Upper swivel joints: This applies to some vehicles fitted with MacPherson strut suspension systems. The upper swivel joints are located on the flitch panel, which is the part of the car under the bonnet beside the wing. With an assistant waggling the wheel, make sure that there is no excessive play in the bearings, and that the flitch panel itself into which the bearing is mounted is not corroded to excess (within 30 cm of the mounting). It should be emphasized here that on most cars the bearing itself cannot be seen, but any wear may be apparent from movement of the suspension mounting.

Note: The condition of the upper swivel joint itself will be further checked when the car is

jacked up and examined from underneath.

Steering rack rubber gaiters: Again, with an assistant waggling the wheel, check that the rubber gaiters are not cracked, broken or perished. The rubber itself is concertina-shaped and the most likely place that it will be damaged is inside the creases so check these very carefully. From under the bonnet only the upper part of the gaiters will be clearly visible, the other side will be inspected from underneath at a later stage. Sometimes severe oil staining provides a clue that the gaiter is damaged.

Steering couplings: On some cars the coupling where the steering shaft is connected to the steering box or rack can be inspected from under the bonnet. If there is a pinch bolt securing the shaft to the coupling check carefully that it is tight, and make sure that the coupling is not excessively worn. This is best checked by seeing if there is any more than perhaps a millimetre of radial 'play' at the outer extremity of the coupling between one side and the other. An assistant waggling the steering wheel helps to show up any wear. At the same time check any pinch bolts clamping the coupling to the shaft (see Figure 4.12).

Steering box and idler arm bearing: On a few cars fitted with this kind of steering mechanism the steering box and idler arm are visible from under the bonnet of the car. If that is the case they will need to be checked at this stage. As far as the steering box is concerned, the most likely cause of wear is in the drop arm bearing (see Figure 4.13). This will be seen easily as the steering wheel is waggled. The shaft attached to the drop arm coming out of the steering box will be seen to move in its housing, and *any* appreciable movement is likely to

Figure 4.12. All steering couplings must be examined. On the Ford Granada shown below, one is located beneath the bonnet (A). It is also vital to check that the pinch bolt (B) which secures the coupling to the steering shaft is done up tightly.

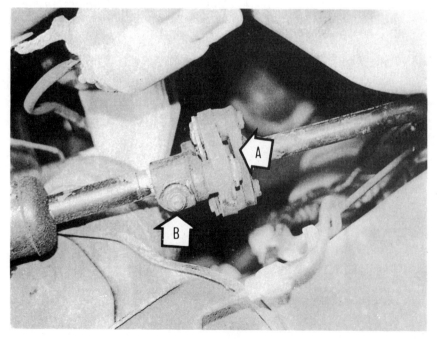

cause your car to fail the test.

On cars fitted with a steering box there is normally a facility to adjust the mechanism to take up the wear. Unfortunately this may not always solve the problem. Sometimes, if the teeth on the gear wheels inside the mechanism have become too worn, they do not mesh properly and the mechanism becomes too tight when the 'play is adjusted from outside.

The steering idler is located on the other side of the car to the steering box, and will have a bearing within which its drop arm shaft rotates. Once more, waggling the steering wheel against the resistance of the tyres will show up wear in that bearing. To tell whether or not the wear is excessive, grasp the drop arm and try to move it up and down by hand. If movement is evident in the bearing, then it would almost certainly fail the MOT.

Corrosion and structural damage: As noted in Chapter 3 there must not be excessive corrosion within 30 cm of any of the mounting points of steering components. Regarding the under-bonnet checks, this will include the steering box, steering rack and idler arm mountings, power steering pump and the upper swivels. The test for corrosion is exactly the same as that explained in Chapter 3.

It should be noted that cracks where the steering box is mounted on the chassis frequently occur, and are difficult to detect. Turning the steering wheel quite hard against the resistance of the tyres on the ground should show up this problem, as the crack will open and close when the steering box tries to move in relation to the chassis.

Power steering: The fluid reservoir will be found under the bonnet, and although this need not be full of fluid to pass the

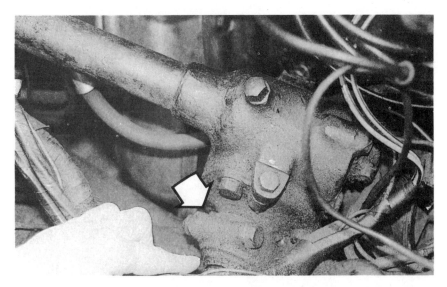

Figure 4.13. Any 'play' in the bearing where the drop arm leaves the steering box could result in a fail. Check this while someone waggles the steering wheel backwards and forwards against the friction of the tyres on the ground.

MOT, if it is almost empty then there is a good chance of a leak in the system somewhere which *would* result in a fail. With the engine running, ask someone to turn the steering wheel from full lock one side to the other and check all the pipes and unions to make sure that there are no fluid leaks (see Figure 4.14). As the steering wheel is turned from lock to lock, the resistance

Figure 4.14. Although the car won't be failed if the power steering fluid reservoir is low, it does indicate the possibility of a leak in the system which would result in a fail. All the pipes and unions must be leak free. On this Ford Granada: (A) fluid feed and return hoses; (B) power steering fluid filler cap; (C) power steering pump unit.

Figure 4.15. For the full pre-MOT test check it will be necessary to raise your car off the ground both with the wheels loaded and unloaded. Sturdy ramps, jack stands and a trolley jack will be essential. Safety precautions are of paramount importance.

should be the same, indicating that the power assistance is operating both ways. If turned on to full lock, most power steering systems will emit a hissing sound as the pressure relief valve opens. This is normal.

Figure 4.16. Whenever the vehicle is jacked up or lifted on ramps be sure to always chock the wheels which remain on the ground as an essential safety precaution.

Under-vehicle inspection

The steering checks underneath your car have to be done with the wheels both on and off the ground. At the testing station these checks will be carried out with the car on a vehicle lift or over an inspection pit, and then using a jacking beam to lift the wheels off the ground later on.

For the average DIY motorist such luxuries are rarely available. The best that can be done is to acquire a pair of small ramps, available at most accessory shops, and drive the front wheels on to them. It is not possible to use jacks and jack stands instead of ramps because it is important at this stage that the wheels cannot easily turn. This enables the various joints to be 'loaded' by turning the steering wheel against the resistance of the road wheels to fully check for wear.

When checks are carried out with the wheels off the ground, then jacks with jack stands will be essential. Because sometimes the steering and suspension has to be loaded during this part of the test, great care is essential. At every stage throughout this part of the test, please be careful.

ESSENTIAL SAFETY PRECAUTIONS

Make sure that the handbrake is firmly applied, that whenever possible the wheels are securely chocked and that the ramps, jacks and jack stands are firmly located on flat and level ground before lying beneath the car when it is off the ground. Be particularly careful when your assistant is applying load to the steering mechanism by either turning the steering wheel, or by pushing and pulling at the road wheels.

The jack supplied with your car for checking the wheels is not acceptable for jacking the car to inspect the underside for MOT purposes. It will be essential to obtain a strong trolley jack and perhaps a ramp. These are available at most car accessory shops (see Figures 4.15, 4.16 and 4.17).

Figure 4.17. Using ramps to raise the car while keeping the suspension under load. Make sure the ramps are of sturdy construction and safe.

It will also be necessary to have some form of *safety* inspection lamp for this part of the examination. There are many such lamps available which operate off the car's cigarette lighter and these are very good and perfectly safe. *Do not use a mains lead lamp because of the danger of electrocution.*

What the MOT test looks for

The main under-vehicle checks on the steering are:

* **Steering swivel joints.**
* **Beam axles.**
* **Steering joints (e.g. tie rod ends).**
* **Steering rack rubber gaiters.**
* **Steering rack or steering box.**
* **Constant velocity joints.**
* **Steering lock stops.**
* **Power steering.**
* **Corrosion and structural damage.**

How to check

You can check these items at the same time as you inspect the suspension joints, with the wheels both on a ramp, and jacked off the ground. Correct jacking when taking the wheels off the ground is vital.

Correct jacking position: To check the swivel joints and (as will be explained later), the suspension joints, your car will need to be off the ramps and have its wheels jacked off the ground leaving about a 2–4 cm gap beneath the tyres. How the wheels are jacked off the ground is vital to properly check the swivel joints on some vehicles, and essential to fully check the suspension on all vehicles.

The MOT Inspection Manual shows how this needs to be done for the popular suspension systems used on modern cars, and this is shown in Figure 4.26 on page 54. The important factor is to jack the car so that the load from the road spring is relieved from the suspension joint or the suspension/swivel joint whichever is the case on the car being inspected.

Swivel joints: With the wheels jacked up, get someone to place a lever (a very large screwdriver or tyre lever) beneath a front tyre and lever it up and down. This has the effect of applying and relieving the load from the various suspension and/or swivel joints (see Figure 4.18). From beneath the car any wear in the upper and lower swivel joint will be evident as relative movement in the joint. Your assistant should now hold a front wheel diagonally and push and pull hard one way and then the other. This should be done diagonally in both ways on each wheel (see Figure 4.19). Again this should highlight wear in the upper and lower swivel joints.

Note: It will be difficult to decide whether there is excessive wear here without some reference to the vehicle manual regarding the design of swivel joint on the car which is being examined. Some vehicles will have spring-loaded joints where movement in the joint is part of the design, whereas others may have solid metal ball joints, where any movement at all (indicating excessive wear) is unacceptable—so check in the manual before examining the swivel joints.

Beam axles: On some vehicles, notably older cars and some commercial vehicles without independently sprung front wheels, there will be a beam axle. In such cases the main swivel joints behind the wheels are the so-called 'king pins and bushes' (see Figure 4.20). These should be checked for wear and end float. With the vehicle jacked up, have someone push and pull at the top and bottom of the wheel, **(without placing their hands or fingers beneath the tyre).** Any wear in the king pin and bushes will be clearly evident. Then ask your helper to lever the tyre up and down. This will identify end-float in the king pin. A small amount of end-float, perhaps up to half a millimetre or so, is acceptable, but if the stub axle clonks up and down, then expect to fail the test!

Steering joints: All of the rods

Figure 4.18. With the wheel correctly jacked up off the ground ask someone to lever the tyre up and down. Check the upper and lower swivel joints as the wheel is being moved. Any wear in the joints can be detected by clear relative movement of the two parts of the joint.

and levers in the steering mechanism (described earlier) will have ball joints or rubber bushes at their ends, and these all need to be checked. Close examination is also required to make sure that all the required split pins and locking nuts are in place. Any additional devices which are sometimes used to transfer the movement will also need to be checked, together with any mountings and bearings which may be used in that particular design of steering system.

Someone's help will be essential here. He or she will need to steadily waggle the steering wheel from side-to-side against the friction of the wheels whilst on the ramp so that all the joints are loaded. Generally excessive wear can be seen as relative movement between connecting components either side of the joint, but sometimes it is best discovered by holding the joint in the hand and 'feeling' for relative movement between the two halves of the joint as the steering mechanism moves slightly from side-to-side. Barely detectable movement is acceptable, but if the movement is very obvious, then the joint is

Figure 4.19. A further check on the swivel/suspension joints can be done by asking someone to grasp the wheel diagonally and to push and pull. This will need to be done across both diagonals. Make sure the fingers are not placed beneath the wheel, and push and pull steadily rather than jerkily so that the vehicle is not dislodged from the jack.

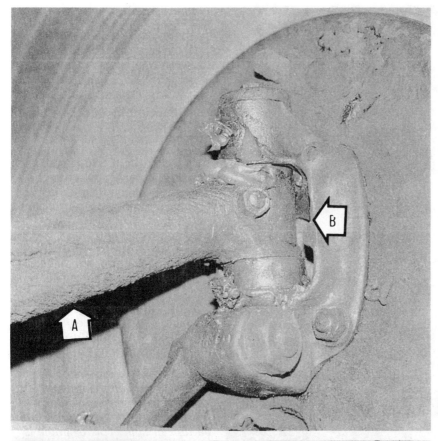

Figure 4.20. With the vehicle jacked up by the axle (A), check the king pin assembly (B) for wear and 'end float'. 'End float' is up and down movement. Side-to-side movement signifies wear in the bushes.

almost certainly worn to excess.

It is worth noting here that on some cars this test will also detect wear in the swivel joint, seen as the joint 'lifting' as the wheels are turned from side-to-side against the friction of the tyres on the ramp. So, it is worth checking while the car is still on its wheels.

Wear in tie rod ends (see Figure 4.21) can sometimes be more easily detected by trying to lift the track rod up and down while the steering mechanism is stationary. A further check will be for your helper to grasp the front wheel in the quarter-to-three position and, with the wheels jacked off the ground, push and pull the wheel one way and then the other against the resistance of the steering mechanism to further highlight excessive wear (see Figure 4.22).

On cars fitted with a steering rack it is sometimes difficult to tell whether apparent wear is caused by a problem in the inner tie rod ends, attached to the rack itself and behind the rack gaiters, or in the outer tie rod ends attached to the stub axle. The difference is important because if the wear is at the inner tie rod ends a new rack could be needed.

Steering rack rubber gaiters: On most cars these will have to

Figure 4.21. Check very carefully for wear in the tie rod end (A). This is a common cause of test failure. The wear is generally evident when someone pushes and pulls the wheel in the quarter-to-three position. It is also advisable to grasp the tie rod (B) near the joint (track rod end) and try to lift it against the joint to detect wear.

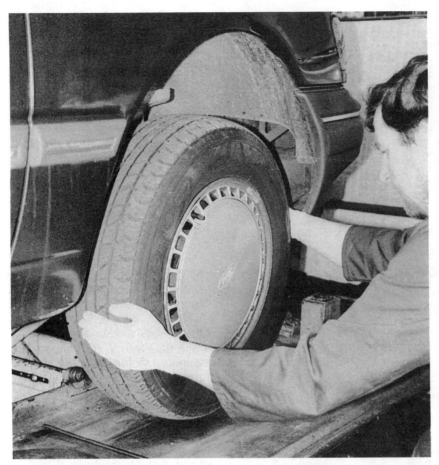

Figure 4.22. To thoroughly check the steering joints ask someone to grasp the wheel in the quarter-to-three position and steadily push and pull. This will highlight any wear in all the steering joints connected to that side. Repeat the process on both front wheels. (Take care not to jerk the car off the jack.)

be checked from underneath the car, and the examination is the same as explained earlier in the section on under-bonnet checks (see Figure 4.23).

Constant velocity (CV) joints: These are the joints behind the front wheels on front wheel drive cars which transfer the power to the road wheels. In practice, unless a joint is severely worn it is difficult to detect CV joint wear during an MOT test.

Figure 4.23. On most cars the steering rack gaiter (arrowed) will be best checked from underneath the car. It should not be torn or split. This example is on a Ford Granada.

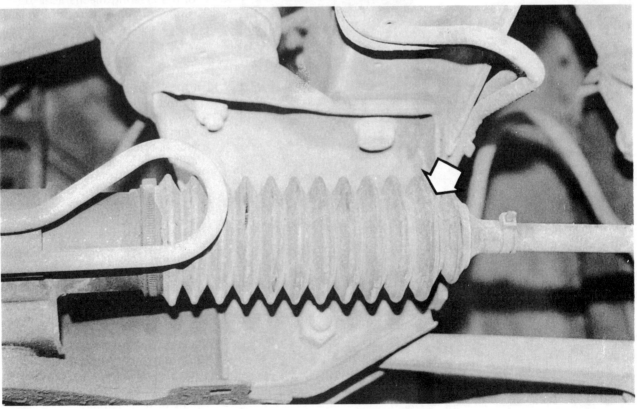

However, the joints invariably have a rubber gaiter acting as a dust cover (similar to the rack gaiter) which has to be tested. It, too, is a concertina-like cover located just behind the wheel and must be examined for tears and holes. As the gaiter turns with the wheel, the testing procedure is to gradually turn the wheel when it is off the ground, and inspect the whole circumference of the gaiter (see Figure 4.24).

Note: Most front wheel drive cars with constant velocity joints located behind the front wheels will also have inner joints taking the drive out of the gearbox. These inner joints are also likely to have rubber boots, but these are *not* subjected to an MOT examination.

Steering rack or steering box: If the rack or box could not be examined from under the bonnet, then it will need to be checked from beneath the car. The examination will be the same, checking for any wear or movement while the steering mechanism is 'loaded' by a helper waggling the wheel. In particular, check where the steering shaft enters the steering box or steering rack housing, and see how much it turns before either the rack itself, or the steering drop arm moves. If the 'slack' movement is excessive, then it will fail the test. As a rule of thumb, if the 'free play' at the wheel was found to be excessive for your car's system when checked from inside the car (earlier), and there is no significant wear in any other joints, then the steering rack or box is almost certainly the cause. There is no laid down rule here, and ultimately it is a matter of judgement for the MOT tester.

Lock stops: With the wheels off the ground ask someone to turn the steering from lock to lock so that the condition and position of the lock stops can be examined.

Figure 4.24. The constant velocity joint is protected from dust and stones by a concertina-like rubber gaiter which turns with the wheel. Inspect the whole circumference for tears. Small holes are acceptable.

It is important here to make sure that the lock stops are there, and have not been adjusted incorrectly so that the wheels or tyres foul the car's bodywork. *Note*: Not all cars have lock stops which can be readily seen.

Power steering: If your car has got power steering, all the under vehicle steering checks will have been completed with the engine running. Your helper should also be asked to turn the steering wheel about a turn one way and then the other. This should highlight any leaks as the power assistance operates. Also, when

Figure 4.25. All the power steering pipes and unions must be free from fluid leaks.

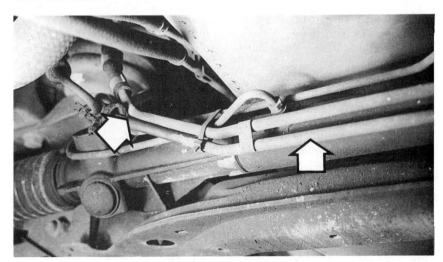

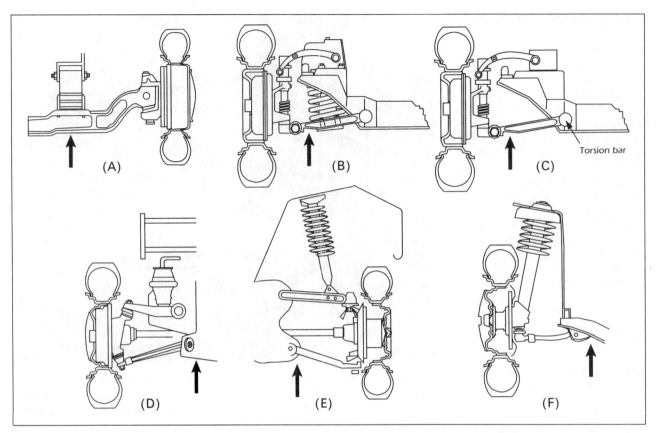

Figure 4.26. Suspension types with jacking points. (A) Front view showing beam axle front suspension with king pins and bushes providing the steering swivel. (B) A system with upper and lower suspension arms, with the lower suspension joint separate from the swivel joint. (C) This system is identical to that shown in (B) except that a torsion bar spring is used (shown here end on). (D) A hydro-elastic suspension system. The drive shaft for the front wheel is shown. (E) Similar mechanics to (D) but the spring, together with a shock absorber, is located beneath the wing with the suspension mountings protruding under the bonnet. (F) This is the classic MacPherson strut system. The upper swivel is at the top of the strut itself and located beneath the front wing with the suspension mounting protruding under the bonnet. The shock absorber is incorporated into the strut.
Note: In (A), (B) and (C) the jack is placed beneath the suspension arms or axle, while in (D), (E) and (F) it is placed beneath the chassis. (Reproduced with the kind permission of the Controller of Her Majesty's Stationery Office.)

the wheels are turned to full lock, if the pressure relief valve fails to operate there could be fluid leaks when the steering is held on full lock. All the pipes, unions and so on should be thoroughly checked for leaks (see Figure 4.25). For reasons of safety, to avoid the vehicle toppling off the ramps when the steering is turned, this check should be done with the wheels firmly on the ground.

Corrosion and structural damage: As with the under-bonnet checks, there must be no excessive corrosion or structural damage within 30 cm of where any steering components are attached to the vehicle structure. Apart from accident damage where the car may have been 'grounded' (over a high kerb perhaps), the most likely damage is from fatigue cracking where the steering box is attached. The method of detection is the same as that described in the section on under-bonnet inspection earlier.

FRONT SUSPENSION

As noted earlier, on some cars the primary front suspension joints and steering swivel joints are the same, although this is not necessarily the case. Figure 4.26 shows the popular types of suspension systems as identified by the Vehicle Inspectorate in their publication *The MOT*

Inspection Manual. However, this does not cover all systems to be found and it is important to ensure whatever system is used on your car, that the load of the car is removed from the joint by appropriate jacking before examination.

JACKING PRECAUTION
It was noted earlier that it is

essential to take great care when jacking up your vehicle for any inspection work to be carried out from beneath the car. This is particularly important when examining the front and rear suspension when the vehicle may have to be pushed and pulled by a helper.

What the MOT test looks for
The main items tested here include all the primary suspension joints, road springs, shock absorbers, suspension tie rods (as opposed to steering tie rods), wheel bearings and the anti-roll bar, if one is fitted, and its bushes.

How to check
The following checks can take place at most stages of the examination of the car, and because of variation in vehicle designs the suspension components will be checked in different places accordingly, but the first step is to check that the shock absorbers are working properly.

Outside the car and underneath the wings
The item being examined here are:

* **Shock absorber operation.**
* **Leaks in the shock absorbers.**

At each corner of the car the bodywork should be 'bounced' by leaning hard on the car so that it settles on its suspension, then letting it bounce back to see if the shock absorber at that corner is working properly. If it is not working, then the car will continue to bounce up and down on its springs for some time. If the shock absorber is doing its job, then the car will stop bouncing very quickly after just one or two 'bounces' as it damps the motion (see Figure 4.27).

On some vehicles the shock absorbers can be seen under the

Figure 4.27. The operation of the shock absorbers is checked by bouncing each corner of the car. It should 'settle back' after just one or, at most, two bounces. If it continues to bounce up and down then there is a problem with the operation of the shock absorber at that wheel.

Figure 4.28. Sometimes the shock absorber is located beneath the front wing, as in the lever arm unit shown here on a Morris Marina. Note the damp area from a very slight leak (A) – this would not necessarily result in a fail. On this design the shock absorber arm also acts as the upper suspension arm (B).

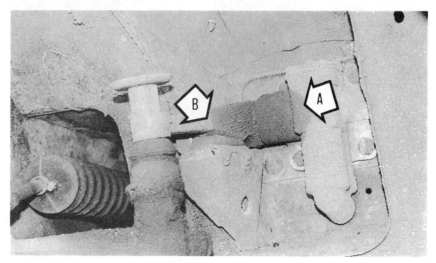

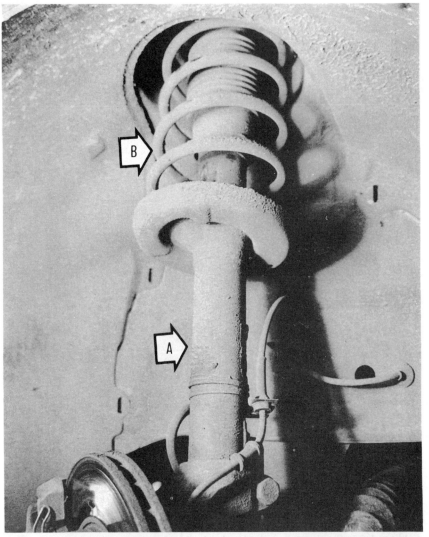

Figure 4.29. A MacPherson strut assembly located beneath the front wing of a Ford Granada. The telescopic shock absorber component should not be leaking. (A) MacPherson strut, (B) road spring.

front and/or rear wings. Have a good look for leaks. If there is a slight leak so that the dust on the shock absorber body is only damp (see Figure 4.28), that is unlikely to fail, but if quite wet, then it will not pass the MOT.

Note: On cars with MacPherson strut front suspension, the shock absorber is built into the strut itself, and can be seen underneath the wing (see Figure 4.29).

Under the bonnet, inside the boot, or inside the car
The items which may be visible to examine here are:

* **The rear shock absorber upper mountings.**
* **Corrosion or damage in the rear shock absorber upper mountings.**
* **The front shock absorber upper mountings.**
* **Corrosion or damage in the front shock absorber upper mountings.**

Starting at the rear of the car, look inside the boot, or inside the vehicle if it is a van or estate and see if the rear upper shock absorber mountings can be seen. They will be about level with the wheels, located more or less above the axle centre line. If so, then make sure that the rubber bushes are sound and not worn or perished so that the shock absorber is insecure in the mounting (see Figure 4.30).

At the same time, examine the surrounding metal to make sure it is not excessively corroded or damaged causing detachment or

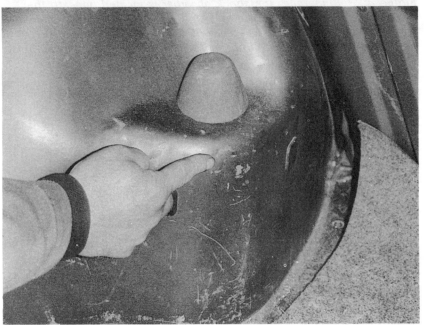

Figure 4.30. If the rear shock absorber upper mountings are accessible from inside the boot they will need to be examined.

insecurity of the shock absorber, although, as was mentioned in Chapter 3, any excessive corrosion or damage in this area may fail the test in any case.

Now look underneath the bonnet. If your car has MacPherson strut front suspension, the upper suspension mounting will have been checked when the steering was examined, since the MacPherson strut is also the shock absorber. Otherwise, there could be a shock absorber mounting which can be inspected from underneath the bonnet, and this must be checked (see Figure 4.31).

Sometimes the shock absorber front mountings are located so they can be seen from under the bonnet. In that case now is the time to check that the rubbers and/or bushes are not excessively worn for the MOT. At the same time have a good look at the surrounding metal to make sure it is not broken or excessively corroded causing it to be either insecure or detached. Again, in the chapter on corrosion it was noted that any problems in the inner flitch area would in any event result in failing the test.

Underneath the car
In this part of the checkover, both the front and rear suspension will need to be examined from beneath the vehicle.

Front suspension: The front suspension is best examined with the wheels jacked off the ground. The precautions regarding safe jacking of the car still apply as noted in the earlier section on checking the steering. The items being checked here are:

* **Upper and lower suspension/swivel joints.**
* **Inner suspension joints.**
* **Beam axles.**
* **Road springs.**
* **Suspension tie rods.**
* **Anti-roll bar and its**

Figure 4.31. Check the front shock absorber upper mountings if they are located under the bonnet. The rubber must not be badly split or perished, and there should be no serious corrosion within 30 cm of the mounting.

bushes.
* **Front wheel bearings.**
* **Shock absorbers.**
* **Surrounding vehicle structure for corrosion or damage.**

It is vital here that your car is jacked up to relieve the load from the main suspension joints, so the jack should be placed exactly as described earlier in the steering section, as shown in Figure 4.26 for the types of suspension illustrated. If your car has a different suspension system, then make sure that when the wheels are off the ground that there is no load on the main suspension joints from either the weight of the car, or the tension in the suspension spring.

Upper and lower suspension/swivel joints: The examination here is exactly the same as that in the section on

Figure 4.32. The front suspension lower joint must be checked for excessive wear. In this case it is not a ball joint but a pin and bush assembly located behind the front wheel. Check the workshop manual to determine the type of suspension system fitted to your car.

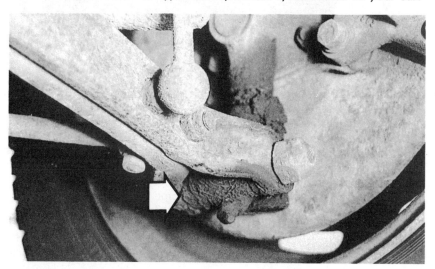

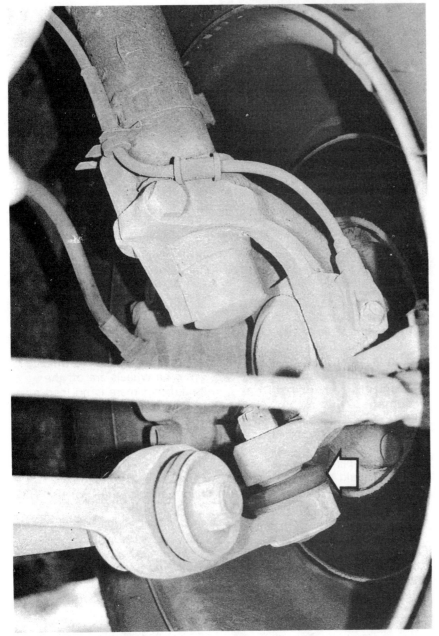

Figure 4.33. The lower suspension and swivel joints shown here are one and the same in the form of a ball and socket.

is vital that your helper relieves all the load at the downward stroke of the lever otherwise the wear will be masked by the spring tension. Check in the workshop manual to see which type of joint is fitted.

Excessive wear in any particular joint can be dependent on the type of suspension system fitted, and reference to the workshop manual will also be essential here. However, unless the joint is spring-loaded with some movement designed into the system, any significant and detectable movement must be suspect.

Sometimes excessive movement can be caused by a worn wheel-bearing which will mask wear in the joint. If the brakes are applied while your helper is levering the wheel, this should lock the bearing, so that any movement left will be caused by wear in the suspension joint.

Now check if the vehicle has an upper suspension joint which will need to be examined, bearing in mind that on some vehicles (with MacPherson struts, for example) the upper joint is located on the flitch beneath the bonnet and was examined earlier.

If your car does have an upper suspension joint which needs to be inspected, again, as the wheel is levered up and down, have a good look for excessive movement in the joint which indicates that it is too worn to merit an MOT pass. The joints on both sides of the car will need to be checked in this way.

Inner suspension joints (upper and lower): On all cars there will be some form of suspension arm or arms (see Figure 4.34) extending from the upper and lower suspension joints connecting the stub axle to the vehicle structure. These, too, will have some form of joint (see Figure 4.35).

If they are rubber bushes, first examine them visually to make sure that the rubber is not split,

the steering swivel joints. However, on some cars the two suspension and swivel joints are not the same, although they could be part of the same assembly of suspension and steering components. In these cases the lower suspension joint itself could be a rubber bush, or a metal pin and bush assembly (see Figure 4.32). It is essential in such instances to make sure

that the bush or bearing is carefully examined for wear.

With a helper levering the wheel up and down, look behind the wheel just below the centre of the hub for the joint itself and check for excessive movement. On some cars where the swivel and suspension joint are one and the same, in the form of a ball and socket (see Figure 4.33), the joint can be spring-loaded, and it

Figure 4.34. The suspension arms connect the stub axle carrying the wheels to the main vehicle structure via the inner and outer suspension joints. All joints must be checked. (A) Front lower inner suspension joint partially hidden from view. (B) Front lower suspension arm. (C) Front lower suspension joint (outer).

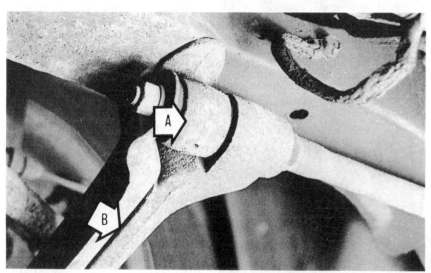

Figure 4.35. The inner suspension joint (shown here on a Morris Ital) will have to be thoroughly examined. Check with the workshop manual for the location and type of joint. (A) Front lower inner suspension joint. (B) Lower suspension arm.

damaged or perished. To check for wear in the bushes or joints it will be necessary to apply a load which pushes and pulls the upper and lower suspension arms in and out. For the DIY mechanic this is not an easy examination. The vigorous pushing and pulling could topple the car off the jack, but it is essential that this check is carried out with the wheels off the ground.

SAFETY PRECAUTION
When carrying out this check make sure that the wheels are only just clear of the ground, and that there are safety jack stands located beneath the vehicle structure if the primary method of jacking is against one of the suspension arms, as will inevitably be the case for some vehicles.

To check these joints, first ask your helper to grasp the lower part of the tyre **(but without placing the fingers directly beneath the tyre)**, and push and pull vigorously but carefully (see Figure 4.36). Look carefully at the lower inner joints to see if there is excessive movement. There could be just one joint or two, depending on whether the vehicle has one or two suspension arms extending from the stub axle. Both must be checked.

Now recheck the safety and

Figure 4.36. With the vehicle just jacked off the ground ask someone to carefully but firmly push and pull the bottom of the tyre. This will highlight wear in the inner lower suspension joints when inspected from beneath the vehicle.

Figure 4.37. In a similar manner to that described in Figure 4.36, ask someone to push and pull carefully from the top of the front wheel. This will highlight wear in the upper inner suspension joints.

Figure 4.38. If the road springs are coiled check that there are no breaks. These can be broken where they bear into the mountings at the top (arrowed) and bottom of the spring. Inspect carefully as any breaks are easy to miss.

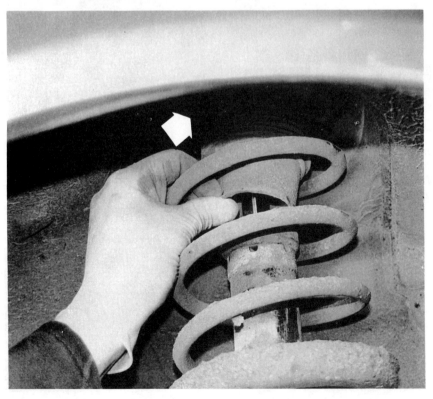

security of the jacks and safety stands, and ask your helper to grasp the top of the tyre and again push and pull very hard (see Figure 4.37). This should indicate wear in the upper inner suspension joint or joints. Both sides of the car will need to be checked.

Now your helper has to grasp the top of the tyre and again push and pull very hard. This should indicate wear in the upper inner suspension joint or joints. Both sides of the car will need to be checked.

Beam axles: As noted earlier in the steering section, some cars will have a beam axle. In that case there will be no suspension arms or suspension joint. This is most likely to be found on commercial vehicles and older cars. The examination is precisely the same as that for the steering swivel joints.

Road springs: The road springs are an important part of the MOT test. In practice there are effectively four different types of suspension springing systems to be found on most types of vehicles. These are: coil springs, leaf springs, torsion bar springs, hydro-elastic suspension, rubber cone (as on the Mini range).

The first thing to check here is if the car sits level on the road. If it leans markedly to one side, then there is an indication that there could be a broken spring, although this is not necessarily the case.

Apart from the hydro-elastic system, it is important to inspect the spring itself, and the mounting. Coil springs are mounted in circular trays into which the top and bottom coils of the spring engage. Check carefully that the part of the spring inside the mounting is not broken (see Figure 4.38). This is often quite difficult to inspect thoroughly and is a common cause of test failures.

Leaf springs (see Figure 4.39)

are more commonly found on vehicles with beam axle front suspension, but a single leaf spring may be used on some older cars where the front wheels have 'independent' suspension. Make sure that all the leaves of the spring are intact, not excessively worn where they are in sliding contact, and properly 'lay' together. It is also vital that any clamping bolts which hold the leaves together are in place and correctly positioned.

Finally, look carefully at the bushes and shackles at the end of the leaf spring unit. Normally these will be rubber bushed. Make sure that the rubber is not split, perished or seriously cracked. On some cars the bushed end of the spring is hung from the chassis with the so-called spring shackles. These must also be carefully examined. Any more than about 0.5 mm of wear would result in a failed test.

If your car has a torsion bar acting as the spring, the bar itself normally lies along the length of the vehicle (see Figure 4.40). If the bar is broken it will be obvious because the car will markedly lean to one side resting on the suspension bump stops (which also have to be examined in the test). Make sure that the ends of the torsion bar are properly secured to the chassis at one end, and to the appropriate suspension arm at the other.

Hydro-elastic suspension systems, and those with only rubber cones providing the springing, do not have a metal spring, but a large rubber 'buffer' arrangement. This is unlikely to fail, and in any event cannot be seen in its entirety to be fully examined in an MOT test. It is worth noting here that if a vehicle with this suspension system leans to one side, then the suspension fluid level on that side needs topping up. This will need to be done with a special pump available at most garages.

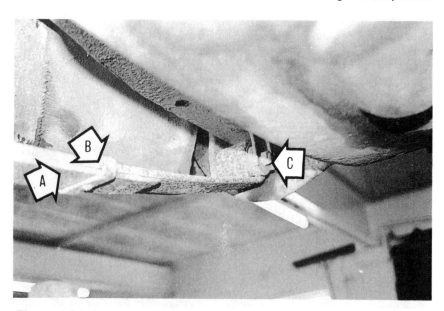

Figure 4.39. Leaf springs (A) will be of similar design and checked in the same way whether they are located at the front or rear of the vehicle, or both. Make sure none of the leaves is broken and that the clamps (B) are properly attached. Check the shackles and bushes (C) at the attachment points at both ends of the spring.

Figure 4.40. Torsion bar suspended vehicles like this Morris Ital will generally have the bar running along the length of the vehicle. Any breakage will be very obvious, but check carefully where the bar is attached to the suspension arm. (A) Torsion bar spring attached to lower suspension arm. (B) Attachment point of torsion bar to suspension arm. (C) Lower suspension arm.

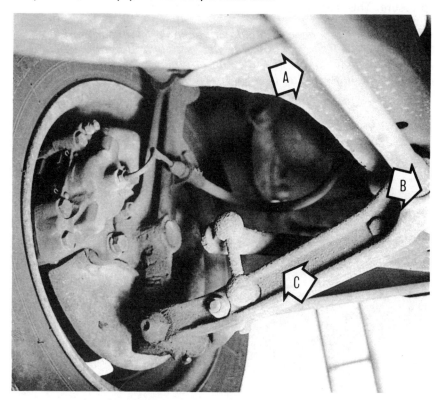

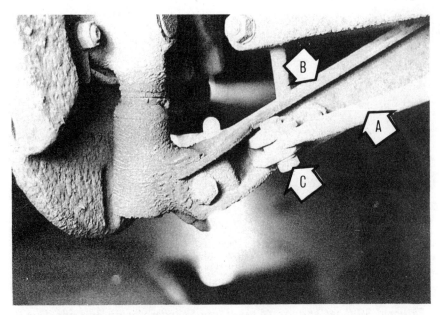

Figure 4.41. The front suspension tie rod (A) is used to stabilize the lower suspension arm (B) if only one arm is used. On this Morris Ital one tie rod joint (C) is attached to the lower suspension arm, and the joint at the other end of the rod is attached to the vehicle structure.

Powered suspension systems: Some cars will have a suspension system which operates when the engine is running and then lifts the car up under hydraulic pressure. This should be examined with the suspension adjusted to the highest 'ride height' setting, and all the hoses, unions and valves should be inspected for leakage.

Figure 4.42. On some cars fitted with a single lower suspension arm (A), called the 'track control arm', this is stabilized by the attachment to it of the anti-roll bar (B), the end of which is joined to the lower suspension arm with the bush (C).

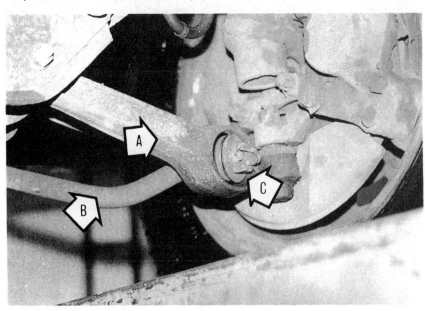

Suspension tie rods: Sometimes the front suspension will have a tie rod from the stub axle going forwards diagonally to the front chassis of the car. This is to stabilize the stub axle and prevent fore and aft movement. The joints at either end of the tie rod (see Figure 4.41) will need to be examined. The first check is to grasp the rod and firmly attempt to push and pull it. Then with the handbrake applied by one helper, ask another to attempt to rotate the front wheel one way and then the other. This has the effect of loading the joints or bushes either end of the tie rod and highlighting wear.

Anti-roll bar and bushes: The anti-roll bar is a 'U'-shaped bar made of spring steel which goes from one side of the car to the other. The centre part of the 'U' is clamped to the chassis on either side of the vehicle, with a rubber mounting, and the ends are attached to the stub axles or sometimes one of the suspension arms on each side. On cars fitted with MacPherson struts and with only a single lower suspension arm, the anti-roll bar is attached to that arm and also acts as a suspension tie rod (see Figure 4.42).

All the attachment brackets and bushes or joints of the anti-roll bar must be carefully examined. First make the usual checks on any rubber bushes for perished rubber, significant cracks and so on. Then the joints need to be loaded to see if there is any significant wear. This is best achieved with the wheels on the ground. With the handbrake applied, ask your helper to attempt to rotate each front wheel forwards and backwards. This will have the effect of pulling and pushing the car against the anti-roll bar and will highlight any movement in the mounting rubbers or attachment bushes.

Movement can be expected in the mounting rubbers which

attach the bar to the chassis on some vehicles, and a reference to the workshop manual would be essential here. Otherwise, any noticeable wear in the mounting rubber would result in failing the test. Where the ends of the anti-roll bar either penetrate the lower suspension arm, or are attached to the stub axle, there should be no movement at all.

If the ends of the anti-roll bar which penetrate the lower suspension arm also act as tie rods, the specially moulded rubber bushes at the point of penetration should be checked carefully. Sometimes, in these cases, significant movement is to be expected and catered for by normal distortion of the rubber in the bush.

Front wheel bearings: These need to be checked with the wheels off the ground, with a helper to lever the wheel up and down. Sometimes it is not clear if movement is caused by wear in the lower suspension joint, or in the wheel bearing. If the movement disappears when the footbrake is applied, then the wheel bearing is likely to be the culprit and excessively worn if the movement is positive and easily detected. Finally, your helper should spin the wheel around. A worn bearing may have no detectable 'play', but wear is indicated if there is a rumbling sound.

On many cars, any 'play' in the bearing can be eliminated by tightening up the nut which secures the bearing. However, be careful and check the manual first. Also, if the bearing itself is no longer serviceable because of wear, even though the 'play' has been removed, it will 'rumble' when rotated after tightening. This indicates that replacement is needed.

Shock absorbers: The shock absorbers could be lever arm or telescopic, but most modern cars are fitted with the telescopic

Figure 4.43. Telescopic shock absorbers will be tested in the same way whether located at the front or (as shown here) at the rear. Check both the upper mounting bush (A) and the lower mounting bush (B) for excessive wear or deterioration, and make sure the unit is not leaking fluid.

type.

Telescopic: (see Figure 4.43). These must be inspected to make sure that there are no excessive leaks. Generally small seepage of the hydraulic fluid is acceptable, and this will be seen as a slight dampness on the surface of the unit, but clear evidence of a leak will result in failing the test.

The telescopic shock absorber will be attached to the vehicle structure at one end, and to the suspension at the other, with rubber bushes. These should not be perished, split or broken to the extent that the eye of the bush has evident movement relative to the shock absorber body. To check these bushes, grasp the main body firmly and push and pull quite hard.

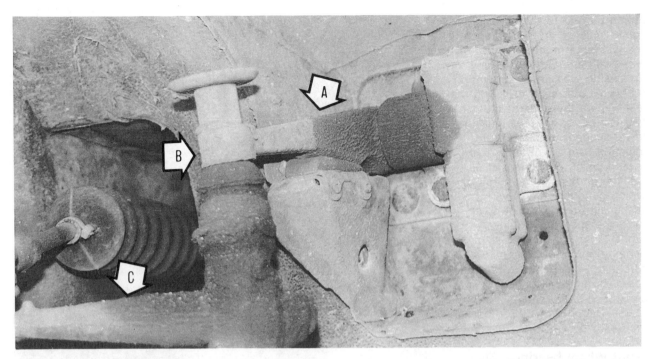

Figure 4.44. On the Morris Ital shown here the shock absorber operating arm (A) also acts as the upper suspension arm with the upper suspension joint (B) also shown. The steering arm (C) also appears in this picture.

Excessive movement of the shock absorber in the bushes is a failure. This is another area where the tester would use judgement, but in practice any more than between 0.5 mm and 1.0 mm of movement could result in a failed test. Another useful check here is to attempt to rotate the body of the shock absorber against the bush. This also highlights any wear.

Lever arm: On some cars with lever arm shock absorbers (see Figure 4.44) the arms constitute the suspension arms themselves and will have already been checked. Otherwise the operating arm from the shock absorber body should be grasped and firmly pulled and pushed to determine whether or not the bushes are worn to excess. These units should also be checked to ensure there are no leaks.

Corrosion and structural damage: As for all mechanical components which are subjected to MOT test there must be no corrosion or damage within

30 cm from where any part of the suspension is mounted on to the main vehicle structure. On many cars corrosion is very likely around the front chassis beneath the front of the car. This is where the anti-roll bar bushes are located, so check this very carefully using the corrosion testing technique explained in Chapter 3.

Note: As a last thorough check have a good look around at all the front suspension joints to make sure that none have been missed. Then, finally, make sure that any split pins, locking nuts and so on throughout the whole steering and suspension system are tight and properly in place.

REAR SUSPENSION

What the MOT test looks for
The items which will be checked on the rear suspension are:

* **All suspension joints and bushes.**

* **Rear road springs.**
* **Suspension tie rods and linkages.**
* **Shock absorbers**
* **Wheel bearings.**
* **Corrosion.**

How to check
The examination of the rear suspension is done first with the wheels on the ramps, and then jacked off the ground taking the same safety precautions as noted for checking the front.

Suspension joints and linkages: The rear suspension systems to be found on most vehicles are very similar to those at the front, but without the swivel joints associated with the steering. The first thing to do is to visually inspect all the suspension components to make sure that there are no obvious problems (see Figure 4.45)—missing nuts, split pins, perished rubber bushes, and so on.

With the wheels on the

Figure 4.45. It is essential to thoroughly test and inspect all the rear suspension joints and linkages. The type of systems can vary greatly from the relatively simple leaf spring/beam axle system shown above (Morris Ital) to the complex independent suspension design with a multitude of joints and linkages shown below (Mercedes 190E).

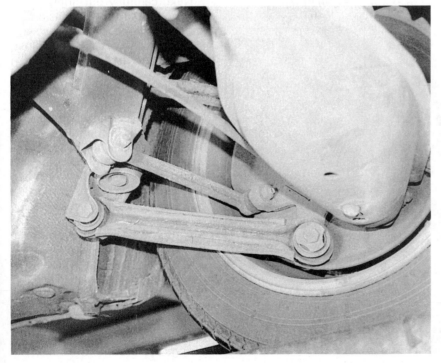

ground, apply the handbrake (provided it operates on the rear wheels), and ask a helper to rock the car backwards and forwards against the resistance of the brake. If the joints and bushes are observed while this is being done, any wear will be detected and likely failure items spotted.

Now jack the car, so that the suspension 'hangs' (see Figure 4.46), **making sure that a safety jack stand or stands are in place as an essential safety precaution, and periodically check that they are in the correct position and that the jack is secure.**

This part of the test must be conducted with great care to make sure that the vehicle does not slip off the jack while your helper is applying load to the suspension system.

With the tyre just a few millimetres off the ground, ask your helper to put a lever beneath the tyre to load the suspension system. While this is being done, have a look at all the

Figure 4.46. The rear suspension must be checked for the MOT test. If possible the rear of the vehicle should be jacked so that the suspension 'hangs', but this is not essential.

Figure 4.47. When checking the rear suspension, ask someone to carefully push and pull the rear wheel in the quarter-to-three position.

joints and linkages to see if any wear is evident. The criteria to apply is the same as that for the front suspension.

Next, carefully but firmly push and pull at all the suspension joints and linkages to detect any movement. Now ask your helper to grasp the road wheel in the quarter-to-three position, and carefully but firmly push and pull (see Figure 4.47). This will highlight any wear which was masked because the joints were under load when the wheels were on the ground. **Once again make sure that hands and fingers are not placed beneath the road wheels.**

Springs: The examination of the rear springs is precisely the same as that for those at the front of the vehicle described earlier.

Suspension tie rods and linkages: In particular when a beam rear axle is fitted, and coil springs are used, and on many

Figure 4.48. *Some cars will have a tie rod linking the rear axle or suspension arm back to the vehicle structure as shown here. This provides greater stability to the sprung road wheel. The tie rod and its ends will need to be examined as part of the MOT test.*

independent suspension systems, some form of tie rods will be fitted to effectively anchor the wheels and prevent 'tramping' around corners (see Figure 4.48). They will all need to be examined for the test. The test here is the same as that for the suspension joints explained above, but inspecting different linkages in the system (see Figure 4.49). Once again, **do not place hands or fingers**

Figure 4.49. *All rear suspension linkages must be checked for the MOT. This link connecting the rear anti-roll bar back to the chassis on a Ford Granada is just one such example.*

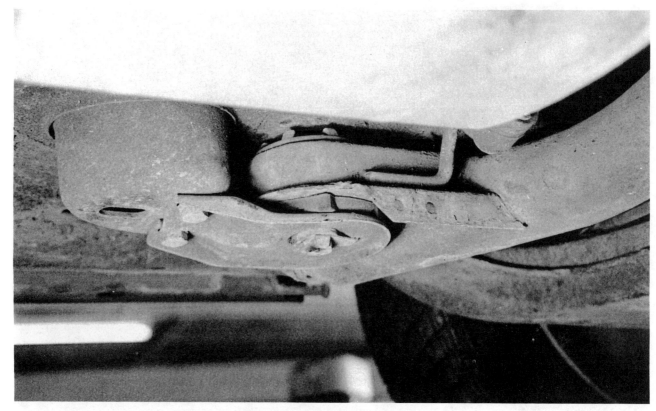

Figure 4.50. On some cars the rear suspension system is mounted on a rear sub-frame assembly which in turn is attached firmly to the vehicle structure. These mounting points must be thoroughly checked for the MOT. Common causes of failure are perished or cracked rubber mounting blocks and structural corrosion where the sub-frame is mounted to the vehicle itself. (The example is a Ford Granada.)

beneath the road wheels while the vehicle is jacked up.

Shock absorbers: The examination of the shock absorbers at the rear of the vehicle will be the same as that carried out at the front.

Corrosion: This will also be identical to the inspection for corrosion carried out on the front suspension, making sure that there is not excessive corrosion or structural damage within 30 cm of where any part of the suspension is attached to the main structure. In particular the attachment points of rear suspension sub-frames can be particularly prone to corrosion (see Figure 4.50).

SAFETY PRECAUTIONS
During all of the suspension checks, both at the front and rear of the car, while your helper is pushing or pulling at the road wheels be very careful that this is not done so vigorously as to topple the vehicle off the ramp or jacks and that when the car is jacked, some form of safety jack or jack stands are in place as a precautionary measure.

QUICK TIPS

Steering

1. Find out if your car has a steering rack or steering box. This is important to check the allowable 'free play' at the steering wheel.

2. By looking at the vehicle, and from reference to the workshop manual, find out what kind of upper and lower steering swivel joints are fitted. If they are ball-and-socket joints, determine if they are spring-loaded, and if any 'play' is allowable. Also, work out which is the correct way to jack the vehicle to examine the lower steering swivel joints.

3. If the outer tie rod ends need to be changed, remember that one side is a right hand thread, and the other is probably a left hand thread. Do not forget to make sure that the 'tracking' is reset. This is not an MOT item, but incorrect adjustment could cause excessive tyre wear, and result in steering difficulties.

Suspension

1. There are a lot of different types of suspension and spring systems. Find out, either from the workshop manual or by inspection, the type of system fitted to your car.

2. Make sure that the vehicle is jacked up so that load is taken off the suspension joints. This will ensure that any wear is readily detected.

3. As for the steering examination, determine if the suspension joints and swivel joints are separate or one and the same. If they are separate, make sure that any apparent movement of the suspension joint indicating wear is not masked or confused with wear in the steering swivel.

4. Make sure that all the joints are loaded whilst being checked so that any wear is clearly highlighted. This means *carefully* but firmly pushing and pulling in all directions.

5. On vehicles with MacPherson strut front suspension, remember that the upper suspension joint is to be found either up inside the wing, and/or beneath the bonnet, and that the shock absorber is an integral part of the suspension leg.

6. Every joint on the rear suspension must be examined, preferably with the load taken off the system. This will normally mean jacking the car so that the suspension 'hangs'.

SAFETY CAUTION: Check the safety notes on Pages 6–8.

Chapter 5

BRAKES

Examination of the brakes has been part of the MOT test since the beginning, and although various makes and models of cars and light commercial vehicles will have different types of braking systems, *every* component connected with the brakes is considered to be a 'testable item' and has to be examined. First it is worth having a quick look at the different types of braking systems likely to be found.

Drum brakes or disc brakes
All vehicles will have either drum or disc operated brakes at the wheels, or a mixture of both.

Drum brakes: In this system, the 'brake drum' is a shallow cylindrical 'baking dish'-shaped component made of steel which rotates with the wheel (see Figure 5.1). Inside it are located two crescent shaped 'shoes', called 'brake shoes' which are attached to the car and unable to rotate (see Fig. 5.2). They have a special friction material attached to their face which is the same circular shape as, and in contact with the brake drum. These are forced against the brake drum either by mechanical means, or by hydraulic pressure acting against a special 'wheel cylinder' (see Figure 5.3) which in turn pushes the shoes against the drum. The pressure of the friction material against the drum slows or stops the wheel from rotating.

Disc brakes: With disc brakes, a specially machined metal disc rotates with the wheel (see Figure 5.4). A 'U'-shaped clamp called the 'brake calliper' (see Figure 5.5) is attached to the car so that the ends of the 'U' enclose part of the disc. The calliper contains metal pads faced with friction material called 'brake pads' which can be forced to come into contact with the disc by hydraulic pressure operating on special cylinders inside the calliper. This contact either slows down or stops the disc, which being attached to the wheel, in turn slows down or stops the wheel. The majority of disc brake systems are hydraulically actuated. A typical disc brake system is shown in Figure 5.6.

Figures 5.1 and 5.2. Drum brakes. The brake drum (A) rotates with the wheel, and the brake shoes (B) are attached to the back plate (C) which is secured to the car. Under hydraulic pressure the wheel cylinder (D) forces the shoes against the drum.

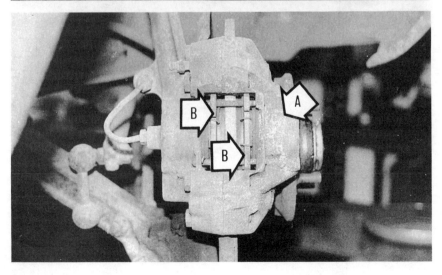

Figure 5.3. **Left.** *Close up of the wheel cylinder (arrowed) which expands under hydraulic pressure forcing the brake shoes against the brake drum to stop the car.*

Figure 5.4. **Middle.** *Disc brakes. The brake disc rotates with the wheel and is gripped by the brake pads.*

Mechanical or hydraulic brakes?

Most vehicles which will be subject to the 'ordinary' MOT test will have either mechanical or hydraulic brakes. Generally modern cars will be fitted with a hydraulic braking system. It is also worth noting that even on cars with hydraulically operated brakes, the parking brake is invariably mechanical. To assist an understanding of the test on brakes for the MOT, it is worth briefly explaining how the different types of braking systems work on cars and light commercial vehicles.

Hydraulic brakes: In this system the brake pedal is connected either directly or indirectly to a piston and cylinder assembly, called the 'master cylinder' which applies pressure to a special oil called 'brake fluid' contained in a reservoir which feeds into the master cylinder. This pressure is carried by metal pipes called 'brake pipes', through special flexible rubber hoses, called 'brake hoses' to either the callipers or wheel cylinders (depending on whether the brakes are disc or drum) at the wheels. The brake hoses have to be flexible to allow the steering to turn, and the suspension to operate whilst still carrying the fluid under pressure when the brakes are applied.

Mechanical brakes: Instead of

Figure 5.5. **Left.** *The brake calliper (A) contains the brake pads (B) which it clamps against the disc with hydraulic pressure.*

Figure 5.6. General view showing a complete disc brake assembly.

using hydraulic pressure to activate the braking mechanism at the wheels, mechanical brakes transfer the force applied at the brake pedal through rods, levers or cables to the brake pads and brake shoes.

Servo assistance: Some cars have a special device which provides additional power to the brakes when they are applied by the driver. This device is called a servo unit, and it has to be examined as part of the test.

The brake servo invariably operates off the suction from the engine intake manifold. This suction works against a piston which either directly assists the effort of the brake pedal, or indirectly introduces additional pressure into the hydraulic system. With such systems the MOT examination of the brakes is conducted with the engine running at idle.

Testing the brakes
The brakes are examined from inside the car, underneath the bonnet and wheel arch, and from beneath the vehicle. Consider first the checks carried out from

inside the car.

Inside the car

SERVICE BRAKE

What the MOT test looks for
This part of the test involves checking:

* Brake pedal and rubber, actuating rods and linkages.
* Brake servo for operation.
* Reserve brake pedal travel.
* Firmness of the brake pedal.
* Excessive corrosion in any visible brake components or the surrounding areas.
* Brake fluid leaks.
* ABS warning system.

How to check
Brake pedal and rubber, actuating rods and linkages:
The first and most obvious check to make is of the brake pedal. It must not be broken or damaged, and the pedal rubber must be in reasonable condition so that the 'tread' on it is not worn smooth (see Figure 5.7). Now, move the brake pedal up and down and side to side by hand to see if there is any side-to-side movement of the pedal, indicating an excessively worn pedal bearing or pivot. Normally there should be hardly any side-to-side movement at all.

Also make sure that the pedal does not foul any other part of the car when it is depressed.

Figure 5.7. If the brake pedal rubber is worn smooth, as in this case, then it will fail the test.

Finally the pedal itself must not be broken, bent or corroded to excess (although that is highly unlikely).

There may be actuating rods from the brake pedal, either to connect it to the main braking system if the vehicle has mechanical brakes, or to the master cylinder if hydraulic brakes are fitted. These must all be checked for wear in any pivots or linkages which can be inspected from inside the car (see Figure 5.8).

Brake servo: To check that the brake servo is operating, first—without the engine running—repeatedly pump the brake to deplete any vacuum remaining in the servo system. Then hold the brake pedal down and start the engine. If some alteration is felt in the position of the pedal, then it can be assumed that the system is working. If the 'feel' of the pedal remains unchanged, then the servo system is probably not working and will require further investigation.

Reserve brake pedal travel: It is important for the MOT test that the brake pedal does not depress very nearly to the floor before the brakes operate. In practice there should be a significant safety margin of reserve travel when the pedal becomes firm and the brakes have been fully applied (see Figure 5.9).

If the brake pedal does depress nearly to the floor before the brakes operate, then your car will fail its test. The problem could be solved by adjustment of the brakes if there are drum brakes in the system (since many cars have a mixture of disc brakes and drum brakes and whilst drum brakes do have an allowance for adjustment either automatically or manually—disc brakes normally cannot be adjusted).

On the other hand the lack of reserve travel could indicate a more serious fault, perhaps fluid

Figure 5.8. The brake pedal (A) may be connected to an actuating rod (B) which operates the piston in the brake master cylinder. The whole assembly will need to be examined. Note the retaining pin locked with a split pin (C).

leaking from part of the system, or a fault within the brake master cylinder.

Firmness of the brake pedal: The next thing to check is the 'feel' underfoot of the brake

Figure 5.9. On checking the 'feel' of the brake pedal make sure it does not depress almost to the floor before becoming firm and operating the brakes. Such 'insufficient reserve travel' is a common reason for failure. Except on a few vehicles where it is normal, the brake pedal should not feel 'spongy' when it has reached the end of its travel.

pedal when it is depressed (see Figure 5.9). As noted earlier, with hydraulic brakes the pedal is connected to a piston which compresses the brake fluid to operate the brakes. If there is air in the hydraulic system, then the pedal does not feel firm when reaching the end of its travel, but is soft and 'spongy'. Such 'sponginess' can result in a failed test, but on some cars this is normal; so if the pedal is slightly 'spongy' but the brakes seem to work perfectly, it is worth checking with the manufacturers—it could be perfectly satisfactory.

Excessive corrosion: It is unlikely that excessive corrosion will be found during the inspection of the brakes from inside the car, but it is worth checking. In particular look at the metal surrounding the area where the pedal is attached to the vehicle if that can be inspected in this part of the check over. It is possible with some cars that water leaking into the inside of the car can cause excessive corrosion within the 30 cm distance laid down.

Brake fluid leaks: Brake fluid leaks are not commonly found inside the car, but are possible on some vehicles when the pedal operates directly onto the master cylinder. In such cases leaks could become evident from inside the car where the brake pedal assembly is attached to the master cylinder through the front bulkhead. Sometimes these areas are not very accessible and cannot be easily seen. It is often easier to 'feel' for fluid leaks by running a finger around the appropriate areas and then checking to see if any brake fluid has been wiped off onto your finger.

ABS warning light system: On vehicles fitted with anti-lock brakes (ABS) the sequence in which the ABS warning light system lights up when the ignition is switched on is part of the MOT test. This is not the same for each type of car. If your vehicle is fitted with ABS, then first consult the vehicle handbook, and if it is not clearly laid out there, contact the local main dealer to find the correct sequence for your car.

PARKING BRAKE

In all vehicles the parking brake operating mechanism is located inside the car, but there are also other parts on the system which will have to be inspected from underneath the car, or perhaps from under the bonnet.
Note: On the majority of cars the parking brake is in the form of a handbrake, but on a few vehicles, notably some American and German vehicles, it is foot operated.

What the MOT test looks for
This part of the test involves checking:

* **The condition of the operating lever and its pivot.**
* **The engagement and operation of the locking pawl, both 'on' and 'off' (the 'clicks' when the parking brake is operated).**
* **That there is reserve travel in the parking brake lever.**
* **Excessive corrosion or damage in a part of the parking brake mechanism or in the surrounding area within 30 cm.**

How to check
Parking brake lever and its pivot: With the parking brake in the 'off' position, check to see how much side play there is in the pivot. If it is excessive, such that the parking pawl could disengage, then it will be failed. On the other hand, this side-to-side movement could be normal for the car. A failure should only apply if the pawl is

likely to disengage *because* of excessive wear in the pivot (see Figure 5.10).

If the lever and pawl mechanism can be inspected from inside the car without dismantling any of the trim, then this must be inspected, together with any locking nuts, split pins and so on.

At this stage the pawl actuating device should be examined. On the majority of vehicles this will be in the form of a button on the top of the brake lever. If the button is missing it will be failed. This is a surprisingly common fault.
Note: On some cars, notably older sports cars, a 'fly off' handbrake is fitted where the pawl operates in the opposite sense in that there are no 'clicks', but when the lever is pulled up the button is pressed to engage the pawl. To disengage the pawl the brake is pulled on that little bit more and the pawl releases. The system must be fully checked regarding condition and operation.

Engagement and operation of the locking pawl: Now the pawl mechanism itself needs to be checked. Allowing the mechanism to 'click', gradually applying the parking brake making sure that each 'click' of the pawl on the ratchet holds the mechanism on (see Figure 5.11). (Obviously this does not apply to the 'fly off' handbrake system).

When the parking brake is fully applied, without touching the pawl operating button or mechanism, knock the top and sides of the lever to see if the parking brake remains 'on'.

Sufficient reserve travel: When the parking brake is engaged it must not reach the end of its working travel, there must be reserve travel to take up wear which may accrue.

If the lever does come to the end of its travel before fully applying the brake, then it will be failed. This could be caused by the need for adjustment of the handbrake system, or of the brakes themselves, or it could be the lever fouling other parts of the car. It is also possible that something may be impeding the movement of the lever. In any event further investigation will be needed.

Corrosion and damage:
Although there is unlikely to be corrosion or damage in the parking brake lever, this must be checked. On the other hand, some vehicles are prone to the metal cracking from fatigue around the handbrake attachment points where the main mechanism is welded to the structure—usually the floor. This is much easier to inspect from beneath the vehicle, but on some cars this can be checked from inside.

To check for cracks or corrosion here if the attachment points can be seen from inside the car, pull the handbrake up hard, and then try to pull it up a bit more. Any problems within the 30 cm distance from the mountings will be clearly evident as the metal moves and the cracks show.

Under the bonnet and beneath the wheel arch
On most vehicles the brake master cylinder and reservoir are to be found under the bonnet, although there are exceptions. The servo unit is also most likely to be located in this area. There will also be various metal brake pipes to be seen from beneath the bonnet and under the wheel arch. The front flexible brake hoses taking the brake fluid to the wheels will also be seen from beneath the wheel arch, except on the rare occasions when there are inboard front disc brakes.

What the MOT test looks for
This part of the test on the brakes involves checking:

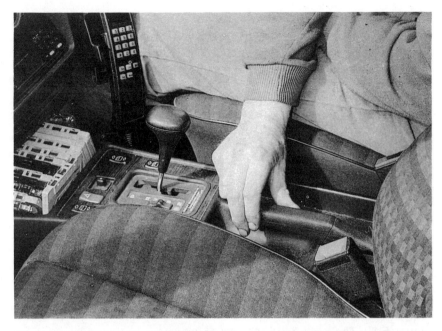

Figure 5.10 Check the side-to-side movement of the handbrake lever when it is in the 'off' position.

* **The brake master cylinder for leaks and general condition.**
* **The brake fluid level in the main reservoir.**
* **Leaks or excessive corrosion in the metal**
  brake pipes and unions.
* **Leaks, tears or excessive deterioration in the flexible brake hoses.**
* **The servo unit, for leaks and general condition.**
* **Corrosion or damage in**

Figure 5.11 Gradually apply the parking brake, checking that each 'click' of the pawl on the ratchet holds the mechanism on. Also make sure that when the brake is fully applied it has not reached the full extent of its travel. Some reserve movement is essential.

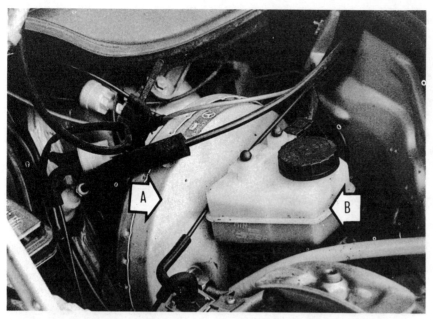

Figure 5.12. The servo unit (A) and brake master cylinder are often connected together to form an integrated unit. Check thoroughly all round for fluid leaks with the brake applied. (B) Master cylinder reservoir.

any braking components or in the areas surrounding mountings.

How to check
Brake master cylinder: This must be checked for any leakage of hydraulic fluid. Often the servo unit is connected directly to the master cylinder and there is a likelihood of leakage where they are connected together (see

Figure 5.13. Check the brake fluid level in the master cylinder reservoir. Not only is a low level a reason for failing the test, but it indicates that there is a leak somewhere in the hydraulic braking system.

Figure 5.12). Ask someone to apply pressure on the brake pedal while carrying out this inspection, and make sure that the engine is running.

Also have a look at any linkages between the footbrake mechanism and the cylinder itself. On some vehicles there is a long linkage across the car from the driver's side to where the master cylinder is located on the passenger's side. Ask your helper to repeatedly operate the brake while you check the linkage for excessive slack and wear.

Note: Sometimes the brake master cylinder is mounted on the chassis and cannot be seen underneath the bonnet.

Brake fluid level: The brake fluid reservoir may be mounted directly onto the brake master cylinder and is often a part of it. Alternatively, it may be located elsewhere under the bonnet with a pipe carrying the fluid to the master cylinder. Unless the reservoir is transparent through which the level of fluid can be seen, remove the cap and check the fluid level (see Figure 5.13).

If the reservoir is remotely located from the master cylinder, carefully examine the pipe between the two to make sure that there are no leaks.

Note: If the fluid level is low your car will fail its test, so be sure to top it up. A low level could indicate a leak of brake fluid somewhere in the hydraulic system.

Condition of brake pipes, unions etc.: There will be metal brake pipes carrying brake fluid from the master cylinder to the various actuating cylinders at the wheels. All these pipes and associated connecting unions located under the bonnet must be examined for excessive chafing or corrosion. Leaks are unacceptable. Ask someone to apply the footbrake to pressurize the system while you check for leaks. If your car has an ABS unit, this will probably be located under the bonnet, and must be checked for leaks (see Figure 5.14).

Whether or not a brake pipe is acceptable to the MOT tester depends on the extent to which its wall thickness has been reduced by chafing, damage or corrosion. If the reduction is more than a 1/3, then it will be failed. Unfortunately this is pretty meaningless, but for all practical purposes if the chafing or corrosion has removed about 1/4 mm of metal, then it will probably fail the MOT.

The clips that retain the metal brake pipes will also have to be inspected because their absence could cause chafing of the pipe, and is a cause of test failure (see Figure 5.15).

Sometimes surface dirt or grit can hinder the assessment of these pipes. This can be removed with a small scraper to properly check the condition of the metal. Finally, make sure that none of the metal pipes is kinked or fouls any moving parts (the road wheel for example).

Flexible brake pipes: Although these will also be examined from underneath the car, they will also need to be checked from beneath the front wings. Ask someone to pressurize the system by applying the footbrake, and carefully examine the outer rubber covering on the hoses to make sure it is not perished, cracked or broken. Note, however, that some slight surface cracking is acceptable. Also examine where the ends are connected by metal unions and washers. There must not be any fluid leaks (see Fig. 5.16).

Make sure that the hoses are not twisted. This is a surprisingly common problem and is caused by incorrect assembly when a flexible hose is replaced. A twisted hose can restrict the flow of brake fluid and will result in an MOT failure.

Figure 5.14. Some cars are fitted with an anti-lock braking system (ABS). The unit is likely to be located beneath the bonnet and must be checked for leaks with the system pressurized.

Figure 5.15. Check all metal brake pipes and unions which can be examined beneath the bonnet. Also make sure they are properly clipped in place.

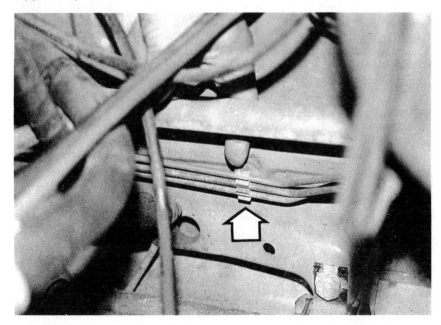

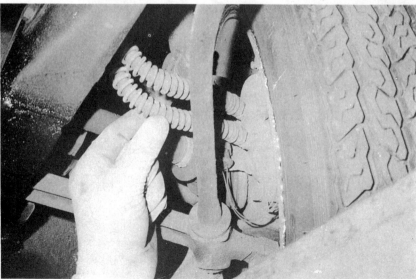

Figure 5.16. Using a low-voltage safety light inspect the flexible brake hoses to be found underneath the front wing.

The last thing to check is that the hoses do not come into contact with the wheels or other moving parts (the brake disc for example). This is normally indicated by scuffing on the surface of the flexible hose itself, but can also be checked by turning the front wheels from lock to lock and having a good look to see if the hoses are always clear of the wheels, discs, drive shafts and so on (see Figure 5.17).

The brake servo: This will almost invariably be located under the bonnet, although on some earlier cars it could be located beneath the front wing. With the engine running, and with someone applying the footbrake, carefully examine the servo unit itself and any nearby pipes and unions for brake fluid leaks.

Normally the servo will have a suction pipe leading to it from the engine of the car. Make sure that the pipe is not cracked or broken, and that the rubber is not perished. A common fault here is for the pipe to collapse inwards when it has deteriorated and the vacuum sucks it in so the opposite sides touch and constrict the pipe. This can also happen if ordinary hose pipe is used to replace the special servo piping. Squeeze the pipe to check that this has not happened (see Figure 5.18). A collapsed pipe preventing the suction from reaching the servo unit will cause your car to fail the test.

Levers, linkages and cables: To some extent this was covered

Figure 5.17. Check the flexible brake hoses with the steering on full lock to make sure they do not rub against any moving parts, drive shafts, the road wheel or brake disc.

in the examination of the master cylinder, but there could be other brake levers and linkages beneath the bonnet, perhaps connected with the handbrake mechanism or the servo unit. These must all be examined to make sure that there is no excessive wear in any joints, frayed cables, or excessive corrosion. Ask someone to work the appropriate mechanisms while you examine them in operation.

Corrosion or damage: The normal inspection for corrosion or damage in any braking components will need to be carried out, as well as a careful look at the metal within 30 cm of the attachment point of any braking component for damage or excessive corrosion. It should be noted that on some cars the under-bonnet area is very prone to rust and corrosion, and this should be very carefully checked.

Underneath the car

To inspect the components of the braking system located beneath the car, it will be necessary to either jack the car up using jacks or jack stands, or to put the vehicle on to special inspection ramps. The same safety considerations apply as have been noted previously, but note here that the parking brake cannot always be used as a safety precaution because it will need to be operated 'on' and 'off' for some aspects of the test of the brakes. The use of chocks to lock the wheels is therefore vital in this case.

It will be assumed in this section that the brakes will be inspected from beneath the car in a methodical manner moving from the front to the rear of the car.

ESSENTIAL SAFETY PRECAUTIONS

* **Chock the rear wheels.**
* **Locate the ramps, jacks and jack stands on firm ground.**
* **Wherever possible use additional safety jack stands.**
* **Use only low voltage safety inspection lamps.**

FRONT BRAKES (HYDRAULIC)

What the MOT test looks for

This part of the test involves checking:-

* **The condition of the brake callipers, brake discs, and pads if disc brakes are fitted (if visible), including checking for leaks.**
* **The condition of the drum brakes as far as can be seen, including checks for leaking wheel cylinders.**
* **The condition of flexible brake hoses as observed from under the car.**
* **The condition of any metal brake pipes and unions and checking for leaks.**
* **The parking brake mechanism if operating on the front wheels.**

How to check

Brake callipers, discs and pads: With someone applying the footbrake to pressurize the hydraulic system, examine the calliper carefully to see if there are any leaks (see Figure 5.5 on page 71). On most cars the metal disc will be easily accessible, and it must be inspected to see if the surface is smooth and not scored or pitted (see Figure 5.4 on page 71). The brake disc is normally located behind the wheels, but some vehicles could have 'inboard' discs located nearer the centre of the car. If the disc is excessively rusty or scored in the part where the brake pad bears on it, then it will fail the MOT. The condition of the friction material on the brake pad should be checked to see whether or not there is plenty of life left in the pad, if it can be seen.

It is also important to check that the pins which lock the brake pads in place are there. These are often left out by DIY mechanics, and their absence will result in a failed test. Also make sure that the nuts and bolts which fasten the calliper to the stub axle are all in place, fully tightened, and have the appropriate locking nuts, split pins or locking washers.

Figure 5.18. Check the suction pipe from the engine to the servo unit. It must be free of leaks and not have deteriorated to the extent that it collapses in on itself.

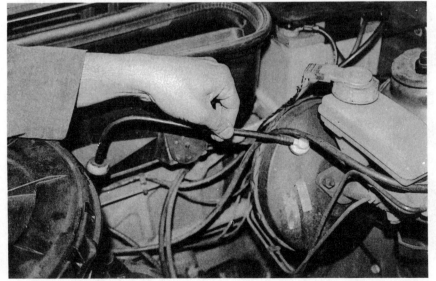

Figure 5.19. If the rear wheel cylinder (A) has a leak this can be detected by brake fluid either staining the back plate (B) or dripping off the bottom of the back plate. That is not the case here. For the DIY mechanic it is worth removing the drums to check for leaking cylinders.

Figure 5.20. Many cars with drum brakes are self-adjusting, but if there are brake adjusters make sure they are not broken, and it is well worthwhile checking the adjustment before taking your car to be tested.

Drum brakes: Although there is little to see if drum brakes are fitted, any internal brake fluid leaks can be detected by a tell-tail stain down the back plate located behind the wheel. Such leaks are normally caused by the wheel cylinders, located inside the drum, leaking due to worn or damaged seals (see Figure 5.19). When checking, make sure that someone is pressurizing the system by applying the foot brake. Bear in mind that the cylinder could be in poor condition and leaking inside the drum, but that the leak is not sufficiently bad to be seen as a stain on the backplate.

It is well worth removing the brake drums as a matter of course when checking prior to an MOT test, even though the testing station will not do so and will have to pass your car on this count if there are no evident leaks. With the drums removed, any slightly leaking wheel cylinders can be detected and remedied before further

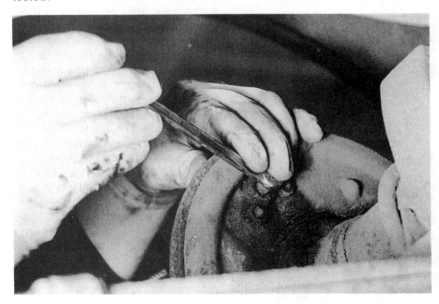

deterioration develops and threatens catastrophic loss of braking from a total loss of brake fluid. Any leaking cylinders should be replaced.

Also located on the back plate on some cars fitted with drum brakes is the brake adjuster (see Fig. 5.20), although many modern cars have so-called 'self adjusting' brakes, whereby any wear in brake shoes is automatically taken up as the brakes are applied. However, if the car does have a brake adjuster make sure that it is not damaged or broken. Again, check that any split pins, locking nuts and other such devices are all securely in place.

Flexible brake hoses: These will already have been checked from beneath the wheel arch, but they may be more easily inspected from beneath the car using the same technique as noted earlier.

Metal brake pipes and unions: With the brakes applied to pressurize the system, check for leaks, chafing, excessive damage or corrosion, as already described.

There will also be metal brake pipes and unions carrying the brake fluid from the front of the vehicle to the brakes at the rear. These, too, will have to be inspected all along the vehicle. The checks will be the same as those already described for metal brake pipes located beneath the bonnet.

Corrosion or damage to the surrounding structure: Now inspect the metal for damage or corrosion within 30 cm of any attachment point of parts of the braking system to the vehicle structure.

Parking brake (front wheel operation): This will only apply on some makes of cars. Check any actuating rods and levers, and all the linkages for excessive wear.

FRONT BRAKES (MECHANICAL)

Although mechanical brakes are rare, older cars may have them fitted. This is just a brief note of guidance for such instances. All the rods and levers will have to be inspected for the test. Look at the condition of the rods and make sure they have not been rubbing against other components of the car. If they have they will fail the MOT test. The yokes and clevis pins and all the connections will need to be examined for excessive play or wear, and all split pins, locking devices and so on must be in place.

Of particular interest here is the adjustment of the brakes. It is essential that all the adjustment—from that of the brake shoes, through the operating rods and cables to the actuating lever behind the brake drum—are completed correctly. Otherwise there is the possibility that the angle of the actuating lever at the brake itself is such that the brakes will not be able to operate through their full travel because the lever moves 'over centre'.

PARKING BRAKE (OPERATING ON REAR WHEELS)

On most vehicles the operating components of the handbrake will be located so that they can only be properly examined from underneath. The mechanism could be by rods or cable, although cable operated handbrakes are more common.

What the MOT test looks for
What has to be tested here is:

* **All rods, levers, locking pins, clevises and so on.**
* **The cables for freedom of movement and lack of fraying or knots (or too many cable 'adjusters').**

* **The brake actuating levers.**
* **The surrounding structure for corrosion or damage.**

Note: The type of brake to be found when the parking brake operates on the rear wheels can vary according to the make of vehicle. It could operate on the same brake shoes and drums as the footbrake, or have separate drums or discs. It is worth checking this in the vehicle workshop manual before inspecting the rear parking brake mechanism.

How to check
Rods levers, locking pins etc.: Where the handbrake mechanism works from underneath the car, visually check all the pivots, levers, joints and so on to ensure that there is not excessive wear, distortion or breakage (see Figure 5.21). Ask someone to repeatedly operate the handbrake while you conduct this part of the inspection.

Cables: The majority of handbrake-operated parking brake mechanisms on cars will use Bowden cables. Again, with someone repeatedly operating the handbrake, examine the cables to make sure that none of the inner cables is seized inside the outer. Then carefully check the full length of any exposed inner cables for signs of fraying (see Figure 5.22).

Cable adjusters: Sometimes special 'adjusters' which clamp onto the cable may have been fitted to the exposed inner cables. Their object is to take up some length of the cable and compensate for any stretching which may have developed over the years on older cars. As a rule of thumb the tester will allow one such adjuster on any one cable, but as there is no laid down criteria it is ultimately up to each tester to decide.

Figure 5.21. The pivots and levers associated with the parking brake mechanism must all be examined. The clevis pin, yoke and split pin shown above (arrowed) will need to be checked.

Figure 5.22. All the operating cables on the parking brake mechanism will have to be thoroughly checked for damage or fraying. With someone repeatedly working the mechanism make sure that none of the cables is seized where there are inner and outer cables.
(A) Operating cable to wheel. (B) Bowden cable from handbrake mechanism (inner and outer).

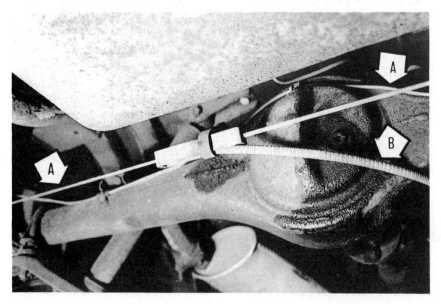

Corrosion or damage to the surrounding structure: Now inspect the metal for damage or corrosion within 30 cm of any attachment point of parts of the braking system to the vehicle structure.

REAR BRAKES

Note: It will be assumed that at this stage all the brake pipes and unions from the front to the rear will have been examined already in the manner previously described in the section on the front brakes (see Figures 5.23 and 5.24).

ESSENTIAL SAFETY PRECAUTION
Take the same care as noted when checking the front wheels.

What the MOT test looks for
The further checks on the rear brakes consists of:

* **Rear brake callipers, discs**

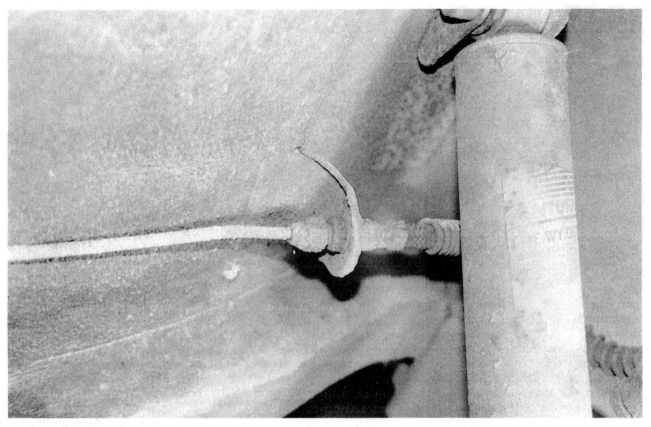

Figure 5.23. Just as at the front, all the rear metal brake pipes and unions will have to be checked for leaks and excessive corrosion or damage.

Figure 5.24. When checking the metal brake pipes make sure they are properly secured with the retaining clips. If they are not, as shown here, they will be failed.

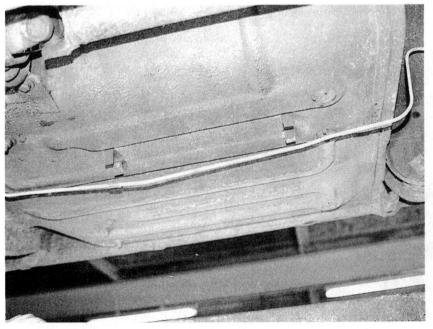

and pads if applicable.
* **Rear brake drums and back plates.**
* **The condition of flexible brake hoses.**
* **The condition of the brake compensating valve (if applicable).**
* **Corrosion and damage.**

How to check
Rear callipers, discs and pads (where applicable): Your car may or may not have rear disc brakes. If disc brakes are fitted to the rear, then the callipers, discs and brake pads must be inspected in the same way as those at the front. It should be noted that the discs are more likely to be located inboard on the rear than on the front.

Brake drums and backplates: The examination here is precisely the same as for the front if your car is fitted with drum brakes all round—checking the back plates for leakage, the adjusters, split pins, locking nuts and so on (see

Figure 5.25. The rear brakes must be checked for the MOT test. Although the tester is not allowed to dismantle anything during his examinations, the DIY mechanic would be well advised to remove rear brake drums to check the condition of the cylinders and shoes. The shoes shown here are in acceptable condition.

Figure 5.26. Check the rear flexible brake hoses for leaks, chafing, cracks or general deterioration of the rubber which may cause 'ballooning'. Ask someone to apply the footbrake to pressurize the system. Make sure all the hoses have been checked. There could be three or more if your car has independent rear suspension.

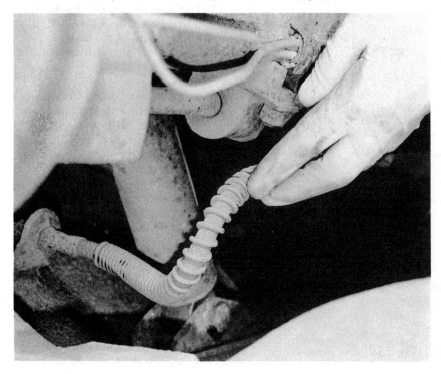

Figure 5.25).

Flexible brake pipes: To allow the rear suspension to operate, flexible brake pipes are essential –to carry the fluid under pressure to the rear wheels. There may be only one, or many flexible brake pipes fitted depending on the design of the rear suspension. All these flexible pipes will need to be checked in the same way as those at the front of the vehicle described earlier in this chapter (see Figure 5.26).

The brake compensating valve: On some cars and light commercial vehicles there is a device within the hydraulic system between the front and rear brakes. It is a regulating valve which alters the relative pressures applied to the front and rear brakes to compensate for the different loads on the front and rear wheels.

Sometimes it is merely a regulating valve, whereas more sophisticated devices are connected to the suspension to change the pressure as the rear suspension moves up and down. In both cases the compensating valve must be examined for the MOT test (see Figure 5.27).

There must be no leaks with the system pressurized, and if there is a mechanism connected to the suspension, then it must be operating freely, and none of the joints should be worn.

BRAKE PERFORMANCE

This is an aspect of the test which the DIY motorist will not be able to duplicate. At the testing station the performance of the footbrake and parking brake are checked accurately at each wheel using the rolling road brake testing machine (see Figure 5.28). The performance obtained is then translated into a percentage efficiency to decide if a pass or fail applies (see Fig. 5.29). It will fail either because

Figure 5.27. Some vehicles will have a brake compensating valve which regulates the braking pressure from front to rear. Sometimes this will be connected between the chassis and suspension to vary the strength of the rear brakes as the rear suspension operates (shown here). Make sure there are no leaks and that the mechanism operates freely.

Figure 5.28. At the MOT testing station the braking performance is checked at each wheel. This is translated into a percentage efficiency to decide if the brakes have passed or failed. The relative balance of the brakes is also checked from side to side.

the efficiency is too low, or because the braking is not even and would cause the vehicle to veer under braking.

Obviously this cannot be duplicated during normal driving by the DIY mechanic. However, there are some simple checks which can be done whilst driving which will indicate if the brake performance is such as to cause the vehicle to veer one way or the other.

ESSENTIAL SAFETY PRECAUTION
If checking the brakes while driving on the road, keep a good look out. Never exceed 20 mph, and only conduct these checks on quiet roads without parked vehicles. If in doubt do not carry out this part of the check.

Footbrake: Drive the car along at a steady speed of about 15–20 mph and, after a careful safety check, gradually apply the footbrake. See if the vehicle veers one way or the other. As the speed drops below 5–10 mph, after a very thorough check to make sure there are no cars

mechanism disengaged to ensure that the brake does not hold 'on'.

Note: Even if these tests indicate that there are no apparent problems, do not be too surprised if your car fails at the testing station, since the specialized testing equipment used is much more sensitive.

Figure 5.29. The brake performance figures are translated into a percentage using a special circular slide rule and from reference to weight charts to determine how much the vehicle weighs.

behind, apply the brake more firmly to see if the car still stops in a straight line. Before the car comes to a halt, release the brake to see if it still travels in a straight line. (*Note:* this last check will indicate any tendency for one of the brakes to 'grab on'.)

It may be necessary to carry out this test a few times to be sure that any problems which it can show up have been identified.

Parking brake: This check

should only be carried out where the car is fitted with a handbrake which can be applied and released using one hand. It would probably not be a practical proposition to attempt this check with a foot operated parking brake because the release mechanism is often remote from the foot operation and could cause some confusion.

Carry out the same check as explained for the footbrake, except in this instance the handbrake is gradually applied and released with the pawl

QUICK TIPS

Brake pedal too low: If the brake pedal is too low, the reason could be:

1. Insufficient brake fluid in the system. Check the fluid reservoir and, if the level is low, look for a leak somewhere in the system.

2. The brakes could be in need of adjustment, or the brake shoes could be worn down. This is more likely to be the case on vehicles which are fitted with drum brakes all round. With guidance from the workshop manual, examine the brakes and change the shoes, or readjust them if the shoes are in good condition.

3. There could be a fault in the brake master cylinder.

The brake pedal is spongy: The reasons for this could be:

1. There is air in the braking system. With reference to the workshop manual, bleed out the braking system.

2. There is a fault in the brake master cylinder.

3. One or more of the flexible brake hoses is perished or soft and is 'ballooning'. (*Note:* this is a less likely fault).

Brake fluid leaks: There are areas which are more likely to be leaking than others, these are:

1. A wheel cylinder leaking. This is very common, and is detected by brake fluid staining the back plate where it has leaked and run down the plate.

2. From the master cylinder. This is likely in two places. The first is where the actuating rod from the pedal linkage enters the master cylinder to actuate the piston. The second area is where the master cylinder joins on to the servo unit when they are in tandem with the servo applying power to the operating rod.

Brake shoes and cylinders, and pads: Although the condition of the shoes, cylinders and pads is inherently part of the MOT test, because the shoes can never be observed without dismantling the brake drum, and sometimes the pads cannot be seen either, then the tester does not always directly examine these items. The DIY mechanic is under no such constraints, and with the guidance of the workshop manual it is well worth while inspecting and servicing the brakes before submitting the vehicle for an MOT test.

Handbrake: It is a myth that the tester counts the 'clicks' when checking the handbrake. He is only concerned that there is reserve travel. This is an easy matter for the DIY mechanic to check and adjust before taking the vehicle for test.

SAFETY CAUTION: Check the safety notes on Pages 6–8.

Chapter 6

TYRES AND WHEELS

TYRES

This part of the test not only checks the condition of the tyres, but the type and size of the tyres fitted to the road wheels is also examined. *Note:* 1. The spare wheel is not part of the test, and neither the wheel nor tyre will need to be examined. 2. There are very specialized tyre requirements for some larger vehicles which can be tested at a Class IV testing station—notably passenger-carrying vehicles and ambulances with up to 12 seats. These requirements will not be covered here.

Tyre size and type

What the MOT test looks for
This part of the examination of the tyres is checking for:

* **Correct tyre size and mixture of sizes.**
* **The correct tyre type.**

How to check
Tyre size: What is important here is that the size of tyres on the same axle are compatible. In the majority of cases this means the tyres on the front wheels must be the same size, and the tyres on the rear wheels must be the same size.

Look for the size marking on the wall of the tyre. It will have something like 165/13, for example. The first number denotes the size of the tyre and must be the same for wheels on the same axle. The second number indicates the size of the wheel, which obviously must also be the same on each side of the car. This check is not quite so simple, as there may be other numbers on the tyres, and these will refer to the 'aspect ratio' or fatness of the tyre. Nevertheless, the best way to be absolutely certain the tyres will not fail the MOT test regarding size is to make sure that there are identically sized tyres on the same axle (see Figure 6.1).

Notes:
1. If the vehicle being tested has twin rear wheels, and one of the wheels is not in contact with the ground, then that is not a reason for an MOT failure on the basis of tyre size because the tyre touching the ground could be over-inflated. Over or underinflated tyres do not justify an MOT failure.
2. Special 'thin' weight and space saving spare tyres will fail if fitted to a road wheel when the car or vehicle is presented for MOT test.

Figure 6.1. The tyre size is shown on the wall of the tyre. Wheels fitted to the same axle must have the same size tyres.

Tyre type: There could be one of three types of tyres fitted to vehicles which require an MOT test. These are:

* **Bias belted-tyres.**
* **Cross-ply tyres.**
* **Radial-ply tyres.**

Tyre type fitted to any one axle: The type of tyre structure must be the same for tyres fitted to the same axle. Thus both front tyres must be of the same construction—that is, bias-belted, cross-ply or radial—as must both rear tyres.

Tyre type fitted to different axles: The requirement of the MOT here depends on whether or not the vehicle has twin wheels on an axle, has three axles or only two.

First, let's look at ordinary cars and light commercial vehicles with just two axles and only single wheels. The best way of looking at this is to note what results in an MOT failure.
A fail will result if:
1. Bias-belted or cross-ply tyres are fitted to the rear, and radial-ply are fitted to the front.
2. Bias-belted tyres are fitted on the front with cross-ply fitted to the rear.

Note: This does *not* apply on vehicles with two axles and twin wheels on the rear, or for those fitted with three axles of which one steers and one drives, the other just 'idling' to carry part of the load. It also does *not* apply to tyres known as 'super single' which have a road contact area which is at least 300 mm wide.

Tyre condition
The whole tyre must be inspected with regard to its condition. A tyre will fail the test if it is worn beyond the legal limit, or has cuts, bulges or damage which seriously affects the condition of its structure (see Figure 6.2).

Figure 6.2. Check that the tyre has sufficient tread in the right places to pass the MOT. If he is in doubt the MOT tester can use the special tyre tread depth gauge shown here. It is a good idea to visit your local tyre stockist for advice before taking your car for the test.

What the MOT test looks for
* The extent of wear.
* Cuts, bulges or damage.

How to check
The extent of wear: This is measured as the 'tread depth' remaining on the surface of the tyre in contact with the road. The regulations which apply to the majority of vehicles require that there is **at least 1.6 mm** of tread around the whole circumference of the tyre, and that the tread remaining covers the central three quarters of the width of the tyre (see Figure 6.3). The 1.6 mm rule applies to vehicles first used after 2 January 1933 which are:

1. Passenger vehicles (for example, cars, motor caravans etc. which have no more than eight passenger seats).
2. Goods and dual purpose vehicles which do not exceed

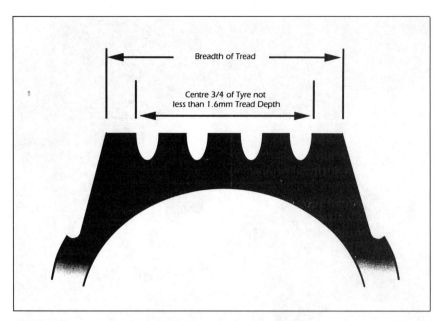

Figure 6.3. Tread requirements. (Reproduced with the kind permission of the Controller of Her Majesty's Stationery Office.)

3,500 kg maximum gross weight.

For other vehicles there need only be 1 mm of tread, these are:

1. Passenger vehicles which have more than eight passenger seats.
2. Vehicles first used before 3 January 1933.

This check is usually visual, but there is a special tread-depth gauge (which is also available to the general public) which can be used if there is any doubt.

Make sure that the parts of the tyre on which the vehicle is resting are also checked by rolling it forward.

Note: For the 1.6 mm check, even if the edges of the tyre are bald, but there is still that central three quarters with adequate tread, then the tyre is adequate for the MOT test. However, if the test is only for 1 mm of tread, then the tread must be visible over the whole tread area.

Cuts: Whether or not a tyre which has a cut will pass or fail the MOT test depends on the length and depth of the cut. The first thing to note is that the cut has to be deep enough to reach the ply or construction cords inside the tyre. Second, a cut will result in an MOT failure if it is 25 mm or longer, or covers more than 10 per cent of the section width (that is the width of the wall or the treaded section, whichever applies).

Bulges (including lumps and tears): The criteria here is whether or not the bulge, lump or tear has been caused by separation or partial failure of the tyre structure, including lifting of the tread rubber.

Note: A tyre will also be failed if any of the ply or construction cording is exposed by, for example, chafing of the side wall.

Tyre fitting
This aspect of the MOT inspection of the tyres concerns how the tyre is fitted to the wheel, and whether or not it fouls other parts of the vehicle.

What the MOT test looks for
The checks look for:

* Correct seating of the tyre on the wheel rim.
* The fitting and condition of the valve stem.
* Fouling of the tyre on other parts of the vehicle.

How to check
Seating of the tyre on the wheel rim: Where the tyre fits on the wheel, inspect the wheel rim all round the circumference, both on the inside and outside of the tyre. Make sure that the tyre is evenly and correctly fitted. This is unlikely to be a problem, but has to be checked.

Valve stem: The requirement here is that the valve stem should not be seriously damaged or incorrectly aligned so that the tyre could suddenly deflate. The absence of a dust cap is not a reason for failing (see Figure 6.4).

The most common fault is where the valve stem is 'cocked over' at an angle because the tyre has moved around the wheel under heavy braking. That *would* be a reason for failing. However, this rarely occurs when modern tyres and fitting techniques are involved.

Fouling: If the tyre is catching or rubbing on any part of the vehicle and such rubbing is not part of the original design, then it would be failed. This is most likely to happen on full lock if at some time the lock stops have been incorrectly adjusted.

Another reason for being failed is if the tyres of twin-wheeled vehicles touch, although sometimes with radial ply tyres they may touch on some twin wheeled commercial vehicles when under full load—in which case the vehicle will not be failed on this account.

WHEELS

The inspection of the road wheels only applies to those

being used, and does not extend to the spare.

What the MOT test looks for
This examination is just visual, and applies to:

* Cracks, damage and distortion.
* Security of attachment of the wheel to the hub.
* Wheel nuts or studs.

Note: The hub cap does not have to be removed to carry out this part of the test, so any wheel nuts concealed behind a hub cap do not have to be inspected.

How to check
Cracks, damage and distortion: The most likely problem here will be at the wheel's rim. Check that there are no severe dents in the rim. Dents can be caused by the car being 'kerbed'. Also spin the wheel when it is off the ground. This will show whether it is seriously distorted. Finally have a good look all round the wheel to make sure that there is no other damage or cracking.

Security of attachment: This is simply to make sure that even if the nuts or studs seem to be tight, that they have not been put in cross-threaded, or with the wheel properly 'seated' on the hub.

Figure 6.4. The valve stem must not be damaged or incorrectly aligned. This could cause dangerous sudden deflation of the tyre.

Figure 6.5. A missing wheel nut will result in your car failing the MOT. The tester is not allowed to remove hub caps to check this.

QUICK TIPS

1. Check that tyres on the same axle, that is both front wheels or both rear wheels are the same type. Make sure that the type of tyre construction for the tyres front to rear complies with the requirement.

2. Make sure the valve is not 'cocked' to one side.

3. Check the whole circumference of the tyres for cuts, bulges, and tyre condition. Don't forget the bit of tyre on which the car is standing, roll the car forward to examine it.

4. Seriously 'kerbed' wheels (dented rims) are likely to fail the test.

Wheel nuts and studs: If they can be seen without having to remove a hub cap, make sure that each road wheel has all the nuts and studs required and that they are tight. It is surprising how many vehicles have one or more wheel nuts or studs missing (see Figure 6.5).

Chapter 7

SEAT BELTS

Vehicle Description

1. Passenger vehicles
- with 4 or more wheels
- with up to 12 passenger seats
- first used on or after 1 January 1965

2. 3-wheeled vehicles
- with an unladen weight over 410Kg first used on or after 1 January 1965, or
- with an unladen weight over 225Kg if first used on or after 1 September 1970

Except vehicles
- less than 410Kg unladen, equipped with a driving seat of a type requiring the driver to sit astride it, and
- constructed or assembled by a person not ordinarily engaged in the trade or business of manufacturing vehicles of this type.

3. Goods vehicles, motor caravans and ambulances
- with an unladen weight **not exceeding 1525Kg**
- first used on or after 1 April 1967

4. Goods vehicles, motor caravans and ambulances
- with a design gross weight **not exceeding 3500Kg**
- first used on or after 1 April 1980

except those first used before 1 April 1982, if they are of a model manufactured before 1 October 1979 with an unladen weight exceeding 1525Kg.

Seat Position

Driver's and "Specified Front Passenger's" Seat (See Note 1 below)	Centre Front Seat	Forward Facing Rear Seats
A. Vehicles first used before 1 April 1981: A belt which restrains the upper part of the body (but need not include a lap belt) for each seat.	No requirement	No requirement
B. Vehicles first used after 31 March 1981: A 3 point (lap/diagonal) belt (see Note 2 below)		

Note 1: The "specified front passenger seat" requiring a seat belt is the seat which is
- Foremost in the vehicle, and
- Furthest from the driver's seat

unless there is a fixed partition separating the passenger seat from a space in front of it which is alongside the driver's seat, eg certain types of taxis, buses etc.

Note 2: '3 point belt' means a seat belt which
- i. restrains the upper and lower parts of the torso
- ii. includes a lap belt
- iii. is anchored at not less than three points, and
- iv. is designed for use by an adult

Figure 7.1 (above). Seat belt requirements for vehicles first used before 1 April 1987. (Reproduced with the kind permission of the Controller of Her Majesty's Stationery Office.)

Seat Position

Vehicle Description	Driver's and "Specified Front Passenger's" Seat (See Note 1, page 4)	Centre Front Seat	Forward Facing Rear Seats

1. Passenger vehicles and dual purpose vehicles with not more than 8 passenger seats.	3 point belts for each seat. (See Note 2, page 4)	3 point belt, lap belt or a disabled person's belt.	A. Vehicles with not more than 2 rear seats: Either
			  i. A 3 point inertia reel belt for at least one seat; or
			  ii. A 3 point belt, lap belt, disabled person's belt or child restraint for each seat.
			B. Vehicles with more than 2 rear seats: Either
			  i. A 3 point inertia reel belt on an outboard seat and a 3 point static or inertia reel belt, lap belt, disabled persons belt or child restraint for at least one other seat; or
			  ii. A static 3 point belt for one seat and a disabled person's belt or child restraint for at least one other seat; or
			  iii. A 3 point belt, lap belt, disabled person's belt or child restraint for each seat.
			See additional information on pages 6 & 7.

2. Goods Vehicles	As above	As above	No requirement
3. Vehicles first used before 1 October 1988 which are: • minibuses with up to 12 passenger seats • motor caravans and ambulances with a design gross weight not exceeding 3500kg	As above	No requirement	No requirement
4. Minibuses, motor caravans and ambulances • with a design gross weight not exceeding 3500Kg • first used after 31 September 1988	As above	3 point belt or a lap belt	No requirement

Vehicles which were first used on or after 1 January 1965 will need to have seat belts fitted, but the specific requirement and which seats require belts will depend on the age, and the type of vehicle in question. As this is a bit complicated, the actual requirement is shown in Figures 7.1 to 7.4 which have been reproduced from the official *MOT Inspection Manual*.

What the MOT test looks for
The checks here are:

* **Security of attachment (including corrosion or damage).**
* **Condition and operation.**

Note: It may be necessary to lift the carpets a little to fully inspect the seat belt mountings, but not to the extent of fully removing them, or 'tearing' any attachment adhesive.

How to check
Security of attachment:
Check the points where the seat belt is attached to the car (see Figure 7.5). This will normally be in three places for the standard lap/diagonal belt. Make sure that all the fixing bolts are secure by inspecting them carefully and pulling the belt against them to check that there is no movement.

On some vehicles the belt is attached to the seat. In such cases not only must the belt be properly fixed to the seat, but the seat must be securely attached to the vehicle structure. Of course, this will be checked when the seat itself is tested.

Now examine the area all around, within 30 cm of the attachment points, for damage and/or excessive corrosion. It

Figure 7.2 (left). Seat belt requirements for vehicles first used after 31 March 1987. (Reproduced with the kind permission of the Controller of Her Majesty's Stationery Office.)

Vehicles first used after 31 March 1987. Forward facing rear seats must have **at least** the type and number of seat belts shown below.

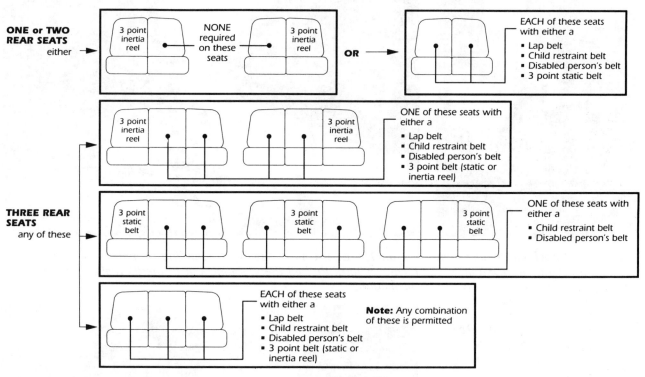

Figure 7.3. Seat belts (additional information: forward facing rear seats). (Reproduced with the kind permission of the Controller of Her Majesty's Stationery Office.)

Figure 7.4. Seat belts (additional information: forward facing rear seats). (Reproduced with the kind permission of the Controller of Her Majesty's Stationery Office.)

should be noted here that on many cars this test will have to be completed from beneath the vehicle to fully examine the mounting points located on the floor. If the vehicle has the belt attached to the seat, then the seat frame must also be checked for damage and/or excessive

MORE THAN THREE REAR SEATS

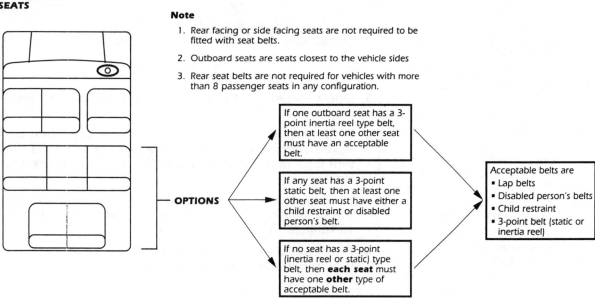

Note

1. Rear facing or side facing seats are not required to be fitted with seat belts.

2. Outboard seats are seats closest to the vehicle sides

3. Rear seat belts are not required for vehicles with more than 8 passenger seats in any configuration.

Figure 7.5. Check all the seat belt attachment points to make sure they are secure, including (as shown here) the inertia reel attachment bolts. The surrounding metal within 30 cm will need to be checked for damage or excessive corrosion.

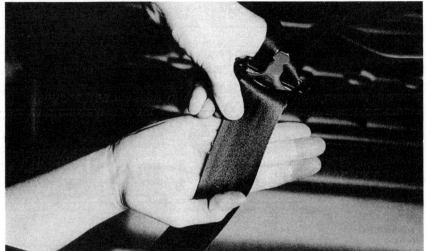

Figure 7.6. Check all the seat belt webbing to make sure it is not frayed or torn. Pull all the webbing out of inertia reels for careful inspection.

corrosion. This check must extend to the surrounding area of the vehicle structure within 30 cm of the seat mountings.

The MOT Inspection Manual notes that on some vehicles it is not possible to inspect the area where the belt is fixed to the seat.

Condition: Carefully examine the webbing of the whole belt, including that inside interia reels (see Figure 7.6). To do this, gradually pull all the webbing out of the reel for inspection. If the webbing is cut or frayed, or the stitching is frayed or damaged to the extent that the strength is obviously weakened, then it will be failed.

The flexible buckle stalks will also need to be examined (see Figure 7.7). Inspect the stalks to make sure there is no damage, corrosion or weakness. Although the metal comprising the stalk is normally covered by a rubber sheath, if the stalk is waggled, any broken strands inside will be indicated by a 'clicking' noise.

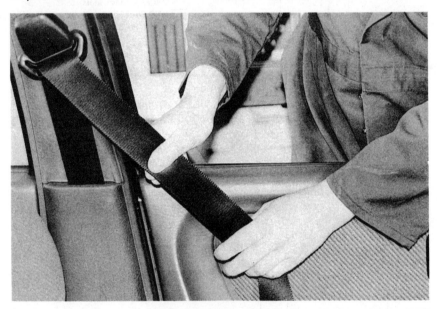

QUICK TIPS

1. **Don't forget to check the webbing inside the reel on inertia reel belts.**

2. **If the belt is attached to the seat, the condition of the seat *structure* will also be checked for the MOT. This will be a more stringent check than that on the seat itself.**

3. **Make sure the clasps lock on *and* release. If the release pin is missing and a finger has to be poked into a hole to release the belt, it will fail.**

4. **From the chart of dates and testable belts, determine which belts are required and have to be tested on your vehicle.**

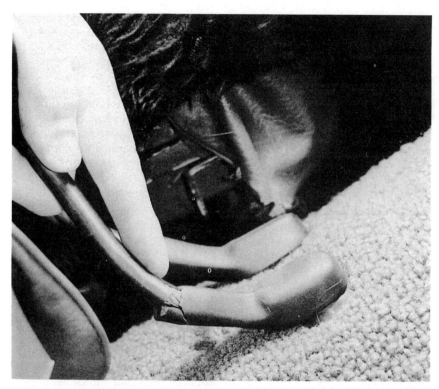

Operation: The locking mechanism will also need to be tested. To check this, first fasten the mechanism, and then try to pull it apart to ensure it has locked properly (see Figure 7.8). Now operate the release mechanism to make sure that the belt releases properly.

If the seat belt has attachment or adjustment fittings, then make sure that they are not fractured or badly deteriorated.

Finally, if the seat belts are of the inertia reel type, gradually pull out the webbing and release it to check that the belt is automatically rewound into the retracting unit. This may have to be pulled back a couple of times before the retract mechanism engages.

Figure 7.7. Examine the condition of any seat belt 'stalks'. The metal cable must not be frayed. Broken rubber covers (seen here) are not a cause of MOT failure.

Figure 7.8. Check the seat belt locking mechanism by engaging the belt into it and try to pull it out. Make sure the release mechanism works properly.

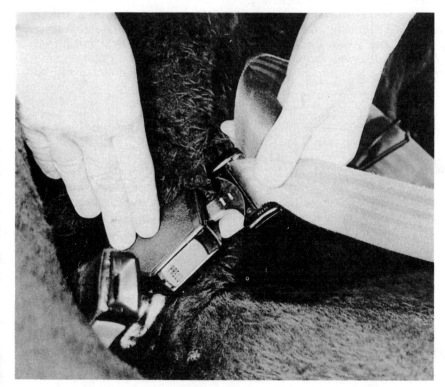

Chapter 8

GENERAL ITEMS

The items covered in this section are:

* **Windscreen, wipers and washers.**

* **Horn.**

* **Mirrors.**

* **Registration plates.**

* **Vehicle identification numbers (VIN).**

* **Exhaust system and exhaust emissions.**

* **Fuel system.**

WINDSCREEN, WIPERS AND WASHERS

Windscreen: If the windscreen has an excessively large crack, star chip or any other damage, then your car will fail the MOT. In practice this depends on how large the damaged area is, and where it is located.

In Figure 8.1 it can be seen that there are two zones on the windscreen covered by the sweep of the windscreen wipers.

Damage in any part of the swept area could result in a fail.

Zone A, as illustrated, is a vertical band in the part of the windscreen swept by the windscreen wiper and centered on the steering wheel, which is 290 mm wide. In this band any damage or obstruction will result in a fail, if it cannot be contained within a 10 mm circle. If there is an obstruction or damage which cannot be contained within a 40 mm circle in the remainder of the screen swept area, it will be failed.

If your car has screen stickers which could result in it failing the MOT, the tester should give the person presenting the vehicle for test the opportunity to remove them before the test is conducted. On the other hand, official stickers such as vehicle licences or parking permits are allowed provided the driver's view is not seriously obstructed (see Figure 8.2). These include parking permits, disabled driver stickers, licenses, and vehicle anti-theft scheme stickers issued by a police authority.

Wipers and washers: The windscreen wipers and washers must work properly and provide the driver with a good enough view of the road ahead to both

Figure 8.1. *Windscreen inspection zones. The requirement regarding windscreen damage or items obscuring the screen is that it must not be greater than 10 mm diameter in zone 'A' centred on the steering wheel, and 40 mm in the rest of the screen area swept by the windscreen wipers. (Reproduced with the kind permission of the Controller of Her Majesty's Stationery Office.)*

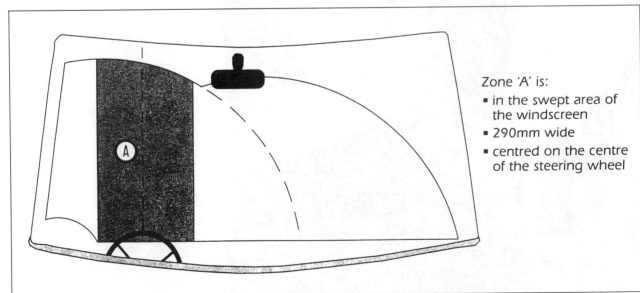

Zone 'A' is:

- in the swept area of the windscreen
- 290mm wide
- centred on the centre of the steering wheel

sides in both wet and dry conditions. In that respect, if only one of the two wipers or washers is working, but the driver's view of the road is still adequate in the way noted, then technically it would pass the MOT. In practical terms, though, the decision is up to the tester. However, to be absolutely sure it is worth checking both wipers and washers.

Wipers: Check that the switch is working properly and turns the wipers on and off. Then, with the wipers working, have a good look at the swept area to be certain that an adequate area of the screen is swept. Sometimes old wipers have worn joints and linkages which result in only part of the screen being swept. If it impairs the driver's view, then it will cause your car to fail the test.

Examine the wiper blades (see Figure 8.3). Old blades could have splits along the length or at the ends, which will cause your car to be failed. Sometimes a fail will result when there are small splits at right angles to the blade length caused by the rubber being perished. New blades are not very expensive and relatively easy to fit, so it would be good practice to change them before

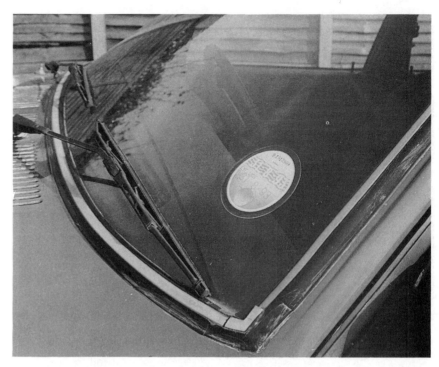

Figure 8.2. Although obstructions on the windscreen outside the laid down limits would result in a failed test, 'official stickers' like the tax disc are allowed.

the MOT test.

Sometimes the wiper arms are not firmly attached to the operating spindles. This could cause an MOT fail, so should be checked.

Washers: If the washers do not provide enough water to adequately clear the screen, first

Figure 8.3. Carefully check the condition of the windscreen wiper blades. The rubber must not be split or extensively perished.

QUICK TIPS

Windscreen
1. 'Official' stickers are allowed—these include the car tax disc.

2. Since this check became part of the MOT test, the windscreen companies have developed specialized methods of affecting acceptable repairs on even quite badly chipped screens. They will offer free advice on whether or not a particular chip or crack would fail the test.

Washers and wipers
1. Before going to the testing station, make sure the washer bottle is full of water. It is surprising how many people forget to check!

2. If there is any sign at all of deterioration of the wiper blade rubber, then fit new blades. They are not expensive, and easy to fit.

Figure 8.4. Check that the windscreen washer bottle is topped up before taking your car to be tested.

check the washer bottle to make sure there is enough water (see Figure 8.4), then, if there is still a problem, see if the jets are blocked and need clearing. This can probably be done by a small pin. A not uncommon problem is a failure of the washer motor. Unless a simple electrical fault can be found a new motor may have to be fitted.

HORN

The test on the horn is to find out if it is loud enough, and has the right tone and if the operating switch works properly.

Loudness: Not only should the horn work, but it must be loud enough to be heard by another road user.

Tone: For vehicles used before 1 August 1973 the only requirement is that the tone must not be an alternating multi-tone sound.

For vehicles first used after 1 August 1973 there is the additional requirement that the horn must have a constant note which is continuous and uniform

and not harsh or grating (whatever that means!). This is all a matter of subjective judgement on the part of the tester.

Note: Vehicles are not allowed to have a 'horn' which is a gong, bell or siren except if it is part of an anti-theft device.

MIRRORS

As in so many aspects of the MOT test, the precise requirement for mirrors depends on the age of the vehicle. If it is a passenger vehicle with seven seats or less and was first used after 1 August 1978, then it must have two mirrors. One of these must be an exterior mirror on the driver's side, whereas the other can be an interior mirror, or an exterior mirror fitted to the passenger's side of the vehicle.

All other vehicles which will be MOT tested need only have one mirror. This could be an exterior mirror, either on the passenger's side, or driver's side, or an interior mirror.

Not only must the vehicle have the correct number of mirrors, but they must not be loose, or be damaged or unable to be adjusted to be clearly visible from the driver's seat (see Figures 8.5 and 8.6).

Whether one or two mirrors are required, the view to the rear must not be seriously impaired. In practice this means that some small damage which does not seriously impair the rear view

Figure 8.5. Any mirror which is compulsorily required must be positioned so the driver has a clear view to the rear of the vehicle. In practice this means that not only must the mirror be unbroken but capable of adjustment.

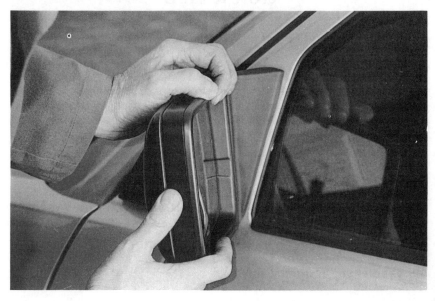

Figure 8.6. This broken mirror prevents the driver from having a clear view to the rear and would result in a fail.

QUICK TIPS

Horn
1. Just make sure the horn works, is loud enough and not a multi-tone of a type which is not allowed.

Mirror
2. Even if the mirror does have a small crack, it may still be acceptable provided there is a clear view through it to the rear. If it is badly cracked, most car accessory shops have 'stick on' mirrors which can be attached over the broken unit.

Figure 8.7. Registration plates are now part of the MOT test, although they would have to be in particularly bad condition (to the extent of being unreadable) to be failed.

would not cause your car to be failed on this account.

REGISTRATION PLATES

If the vehicle is not registered, it does not have to have registration plates for the MOT test.

Security and readability: Otherwise all vehicles must have front and rear registration plates which are securely attached, unbroken and complete. They must also be clean and not defaced, dirty, deteriorated, or obscured (for example by a tow bar), the criteria being that they can be easily read from a reasonable distance of about 20 metres from the vehicle (see Figure 8.7). The number plates must also be unobscured (by a tow bar, for example).

Numbers and letters: The numbers and letters must be laid out in a conventional manner with the spacings between the groups of numbers and letters. If the spacing is altered, or fixing bolts are placed to dot an 'i' in an attempt to personalize plates, then your car will be failed.

Figure 8.8. The vehicle identification number is usually found on a plate like this located beneath the bonnet, although there may be another located elsewhere.

QUICK TIPS

Registration plates and identification numbers
1. **If the registration plate is damaged but clearly readable, then it is acceptable. If it is broken with a piece missing containing a number or letter then it will fail.**
2. **'Stick on' number plates are probably acceptable provided they can be seen clearly.**
3. **Vehicle Identification Numbers (VIN) are normally under the bonnet, but also have a look inside the vehicle around the sill area on the driver's side where they may also be found.**

VEHICLE IDENTIFICATION NUMBERS (VIN)

All vehicles first used on or after 1 August 1980 require a VIN or chassis number, with the exception of kit cars and amateur built vehicles.

The number can be on a separate plate, or it can be stamped or etched onto the vehicle body or chassis. It must be displayed and legible. This is most commonly found under the bonnet (see Figure 8.8), although some cars could have a further plate located elsewhere.

EXHAUST SYSTEM AND EXHAUST EMISSIONS

Although in testing terms the condition of the exhaust system and the exhaust emissions are separate items, in practice the condition of the exhaust system can affect the outcome of the emission test and cause some confusing complications.

If the exhaust system is holed, then the emission levels will be affected because the chemical make up of the gases will be changed. Even so, the emission test still has to be carried out, and the car could pass. Then later, with the the exhaust problem remedied, the emission test could result in a failure because the gases are now

different at the tailpipe where the emission test takes place.

What the MOT test looks for
The main areas to be tested here are:

* **The condition of the exhaust system.**
* **Leaks in the exhaust system.**
* **The noise level from the exhaust.**
* **Exhaust emissions.**

How to check
Condition of the exhaust system: This part of the test will normally have to be conducted

from beneath the vehicle, and the safety warnings noted in earlier chapters will apply with regard to jacking up the car, or driving it onto ramps. Also be aware that exhaust gases are poisonous. Do not run the engine for long periods or in a confined space.

Carefully inspect the whole of the exhaust system (see Figure 8.9). There must be no obvious problems with any part of the system. Neither pipes nor silencers should be corroded to excess with holes apparent (see Figure 8.10), and there should be no seriously deteriorated or missing mountings (see Figure 8.11). On many vehicles the mountings will be rubber bonded, and any delamination of the rubber would cause the tester to fail your car if the effectiveness of the mounting is seriously reduced.

Leaks: This part of the exhaust check has to be done with the engine running. Although small 'pin hole' leaks will be hard to detect, and would not result in a failed test, 'pepper pot' leaks from seriously corroded pipes or silencers are more easy to see

Figure 8.9. The whole exhaust system must be examined, from the engine to the tailpipe.

Figure 8.10. *The exhaust system must be leak-free with all the clamps and mountings secure. Quite severe corrosion is acceptable (as shown here around the clamp and silencer outlet) provided there are no holes and the system is securely attached. (A) Exhaust piping. (B) Clamp. (C) Silencer.*

Figure 8.11. *The mountings which attach the exhaust system to the vehicle must be secure and unbroken.*

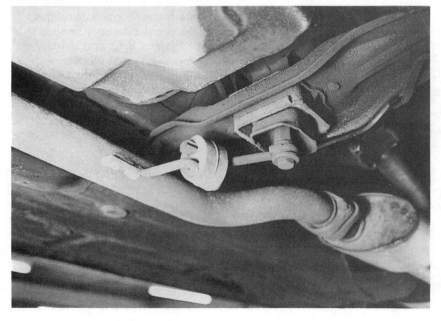

and would cause your car to fail. Although the tester is not allowed to partially block the tailpipe and slightly pressurize the system to aid the detection of leaks, the DIY mechanic is working under no such constraints, and this may well be worth doing to find those small pin holes.

A very common problem is a leak at the exhaust flange connecting the exhaust system to the engine. This is often first indicated by a 'chuffing' sound from inside the bonnet when the engine is running (see Figure 8.12).

Provided that exhaust repairs are reasonably durable and effectively seal the leak then they are acceptable for the MOT. This means that exhaust repair paste can be used on small leaks.

Noise levels: Sometimes silencers can corrode internally so that the baffles are less effective and the exhaust note becomes too loud. This is an entirely

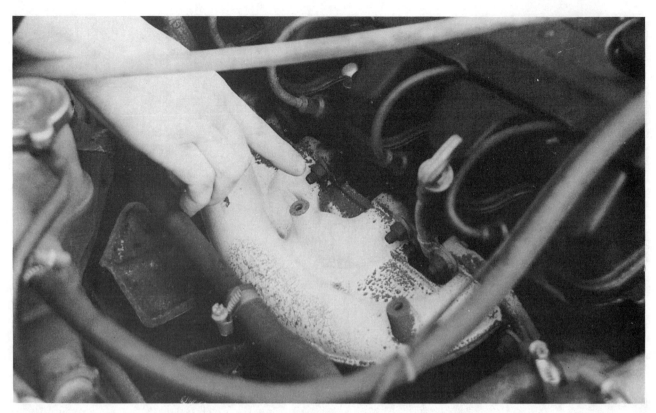

Figure 8.12. *A common problem is an exhaust leak where the exhaust manifold bolts onto the engine. This is almost impossible to remedy without removing the manifold and replacing the gasket.*

Figure 8.13. *During the test the emissions from petrol-engined vehicles are checked with a special gas analyser which measures the level of carbon monoxide and hydrocarbons in the exhaust gases at the tailpipe.*

subjective test on the part of the tester with the criteria being that the noise must not be clearly greater than would be reasonably expected from that kind of vehicle in average conditions.

Exhaust emissions: The first thing to note here is that this is an aspect of the test that the DIY motorist will be quite unable to check because specialized gas analysis equipment is needed (see Figures 8.13 and 8.14). Nevertheless, it is worth going through the regulations and explaining what is required. The requirement is different depending on whether the vehicle is fitted with a diesel or petrol engine.

Dense smoke: All vehicles will have to be checked for emission of dense blue or black smoke. The procedure is specifically laid down. First rev the engine to about 2500 rpm (or about half maximum revs) for 20 seconds and then allow it to return to idle and check for the dense smoke from the exhaust. The emission

Figure 8.14. *The level of carbon monoxide and hydrocarbon emissions in the exhaust gases is measured at the tailpipe by a special probe which sucks them into the gas analyser.*

of such smoke is a reason for failure.

Note: If the engine will not run at a reasonable idling speed, operating at too many rpm, then it will be failed. But if the throttle has to be applied to stop the engine from stalling (that is, the idling speed is too low) it won't result in a fail.

Exhaust emission analysis (petrol engines): Not all cars and light commercial vehicles have to be tested for exhaust emissions with the gas analyser. This aspect of the test only applies to vehicles which were first used on or after 1 August 1975. Some other types are also treated as if they were first used *before* 1 August 1975. These are kit cars, amateur-built vehicles and Wankel rotory engined vehicles first used before August 1987.

Emission standards: The acceptable level of emissions depends on the age of the vehicle. The test levels required are:

1. *If the vehicle was first used on or after 1 August 1975, but before 1 August 1983:*

Hydrocarbon content: The hydrocarbon content of the exhaust gases must not exceed 1200 parts per million for a continuous period of five seconds.

Carbon monoxide content: The carbon monoxide content of the exhaust gases must not exceed 6 per cent for a continuous period of five seconds.

2. *If the vehicle was first used on or after 1 August 1983.*

Hydrocarbon content: The hydrocarbon content must not exceed 1200 parts per million for a continuous period of five seconds.

Carbon monoxide content: The carbon monoxide content must not exceed 4.5 per cent for a continuous period of five seconds.

Diesel-engined vehicles: The test

QUICK TIPS

1. **Make sure that the exhaust is not leaking, repair paste is normally acceptable. If a leak in the exhaust system is mended, previously acceptable emissions may then result in a fail on retest—very inconvenient!**

2. **All the exhaust clamps and retaining clips, rubbers etc. must be in place.**

SAFETY CAUTION: Check the safety notes on Pages 6–8.

here will have to be carried out at the testing station which has special equipment to measure the optical density of the emissions. As with the emissions from petrol engines, this cannot be done by the DIY mechanic.

FUEL SYSTEM

The fuel system will need to be checked for leaks from the engine to the petrol filler cap. All this will need to be done both with and without the engine running.

ESSENTIAL SAFETY PRECAUTION
**Be very careful when carrying out this part of the examination.
Petrol is a highly inflammable and sometimes explosive chemical. It can also be corrosive if spilled onto the hands. It is a heavy vapour. Take great care if using an inspection pit.**

Note: Some parts of the fuel system may not be readily accessible for inspection. If that is the case, it must be assumed that they are satisfactory.

What the MOT test looks for
The specific items which need to be checked here are:

* **The fuel tank.**
* **Visible fuel hoses, pipes and unions (including the carburettor and pump).**
* **Fuel cap.**

How to check
The fuel tank: This will need to be inspected both from above (perhaps inside the boot), and beneath the car. Have a good look all round those parts of the fuel tank which can be seen without dismantling any components. Look for staining or dampening of any dust and dirt which is the tell-tale sign of petrol leaking. Also check for any evidence of petrol smells. In

---

QUICK TIPS

1. Remember the *whole* system has to be checked from the filler cap to the carburettor (if fitted), and that a failure will require the whole test to be completed again, even if the problem was only the filler cap. So check the cap before taking the car for test.

2. Don't forget that some cars will have fuel pumps and filters located underneath the rear of the vehicle beneath the petrol tank. These have to be checked for the test.

SAFETY CAUTION: Check the safety notes on Pages 6–8.

particular check around the pipe which leaves the tank, if it is visible.

Hoses, pipes and unions (including the carburettor): Moving from the engine along the length of the car examine the whole fuel system, starting with the engine compartment. Examine the carburettor with the engine running to see if it has any leaks. Sometimes, when the leak is not very extensive, the petrol evaporates before making a stain. Checking the carburettor with the ignition switched on but the engine NOT running will reveal any such small leaks in the carburettor itself. This will only apply if the vehicle has an electric petrol pump.

Now trace the petrol piping back from the carburettor to the petrol tank. Make sure that all the fuel system components are secure, including the pipes, unions and various attachment brackets. Also make sure that there are no leaks from any of the pipes, unions or flexible hoses (which are sometimes to be found in the fuel system).

Fuel pumps and filters: The fuel pump and any associated filters which can be seen must be checked to make sure that there are no leaks. The pump could be located in numerous places depending on the type and model of car, but if it can be

Figure 8.15. All components in the fuel system will need to be examined for leaks, including pumps (A) and filters (B). Make sure the engine is running to pressurize the system.

Figure 8.16. The fuel tank cap must be secure and not leak. 'Temporary' fuel caps are not acceptable.

seen, then it must be examined (see Figure 8.15).

Fuel tank cap: The cap must be a positive fit, and must properly seal onto the tank with a sealing washer (see Figure 8.16). The criteria here is that the washer must not be torn, deteriorated or damaged, and that the mounting flange must not be damaged such that fuel is likely to escape, for example, when cornering.

Note: The 'temporary' fuel caps which can be purchased in accessory shops are not good enough for MOT purposes.

Chapter 9

THE TESTING STATION

Just as the MOT test is highly formalized, so is the testing station where the test is conducted. There are quite precise rules which govern the nature of the testing facility, and these are laid down in detail by the testing authorities. The space available has to comply to minimum dimensions; there must be a place from which the test can be viewed by the vehicle owner; and the equipment used has to have approval by the authorities for use by the testing station within the context of the class of vehicles the station is permitted to test. This is explained later when the type of 'authorization' of the testing station is discussed. Every testing station will have its own number which must be shown on both the pass and failure certificates it issues.

FACILITIES AND EQUIPMENT

In broad terms, a testing station will have to have the following equipment available for use to carry out a test:

1. An approved roller brake testing machine.
2. A headlamp tester.
3. Exhaust emission testers (petrol/diesel engines).
4. A jacking beam of a size and weight limit appropriate to the testing station's authority.
5. Either an approved vehicle lift fitted with captive turning

plates, or an inspection pit of appropriate size, also fitted with captive turning plates.
6. A small lever or bar to apply load to the suspension system.
7. An approved tyre depth gauge.
8. A small 'official' tapping and scraping 'hammer'.
9. An approved portable decelerometer for use if the roller brake machine fails.

The brake testing equipment and the emission testing machines have to be properly calibrated and periodically checked for calibration in accordance with the testing regulations.

THE MOT TESTERS AND THE AUTHORIZED EXAMINER

Every testing station has an authorized examiner. This is the person or organization, appointed by the testing authorities as being suitable to take on the ultimate legal responsibility for the conduct of the MOT tests at that station. Each testing station will only have one authorized examiner.

The MOT testers are the motor mechanics who conduct the actual MOT test. Until recently they did not require any formal qualifications to become testers. All that was required was to have some years of professional

experience repairing and working on cars, and to have attended a one-day course run by the authorities. When, after a very rudimentary test, they felt that the person concerned was capable of testing vehicles effectively, they issued the person with a tester's number.

Now, to become a tester it is necessary to have certain qualifications in motor vehicle maintenance and repair, or to pass a test set before attendance on the official course. A testing station may have any number of qualified MOT testers on the staff.

The positions of the authorized examiner and MOT tester are held at the discretion of the testing authority, who can withdraw them if they feel it is appropriate should there be a serious breach of the testing regulations.

THE TESTER'S MANUAL

The tester is required to test vehicles in accordance with the procedures laid down in *The MOT Inspection Manual* which is produced by the Vehicle Inspectorate. This is the 'bible' for the MOT test, and a copy has to be available for testers at all MOT testing stations. It has recently been revised and is now somewhat easier to read, but is still essentially a technical document for testing personnel. In the event of any difficulties or ambiguities this is the definitive reference and it has been used as the main source of technical data for this book.

As the reader will already have seen during the discussion earlier on how the test is carried out, some items are clearly of a pass or fail nature, but other items will depend on the judgement of the tester as to whether or not a pass or failure is appropriate.

The tester's manual identifies three criteria for failing testable items as:

1. If a testable item has worn so as to affect adversely the roadworthiness of the vehicle.
2. If a testable item is clearly in need of replacement or adjustment.
3. If a testable item is in a condition which appears to break the law. (Tyres are a good example here.)

THE MOT TESTING GUIDE

The MOT Testing Guide, also published by the Vehicle Inspectorate, provides an insight into the legislative background of the test. It is more important to authorized examiners than to the general public in that it clearly sets out their responsibilities for the conduct of the test at the testing station. It also examines the administration of the test, the day-to-day practicalities of the dialogue between the testing station and the Vehicle Inspectorate, and the various situations in which the authorization held by a testing station can be withdrawn by the authorities.

AUTHORIZATION

Which testing stations can test which type of vehicles?
Not *all* MOT garages can test *all* vehicles which require an MOT. There are different levels of authorization which govern different classes of vehicle, from motor cycles up to large passenger service vehicles. These are shown above right.
All larger commercial vehicles must be tested by the Department of Transport. It is worth checking in advance that a testing station is authorized to test your particular vehicle.

When a test can be refused
It is part of the conditions of authorization that a testing station *must* test a vehicle presented for test which falls

Testing class	Vehicle	Age test first required
I	Motor cycles to 200cc.	3 years
II	All motor cycles (incl. I).	3 years
III	3-wheelers up to 450 kg unladen.	3 years
IV	Cars. Goods vehicles to 3000 kg design gross weight (DGW). Minibuses up to 8 passenger seats. Motor caravans & dual purpose vehicles.	3 years
	Minibuses 9–12 passenger seats. Taxis and ambulances to 12 passenger seats.	1 year
V	Private passenger vehicles and ambulances with more than 12 passenger seats.	1 year
VI	Public service vehicles. These must be tested by the Department of Transport.	1 year
VII	Goods vehicles over 3000 kg and up to 3500 kg DGW.	3 years

within its class of authorization, but there are exceptions when an MOT tester can refuse to undertake the test. These are:

1. If the registration document is not available and it is required to determine the date of first registration to carry out the test properly.
2. If the vehicle cannot be driven and so the test cannot be completed.
3. If a vehicle is so dirty that it is too difficult to test it.
4. A vehicle which has an insecure load, as decided by the authorized examiner, unless the load is either rendered safe, or removed.
5. If a vehicle is too large or too heavy to be safely or properly tested on the approved facilities.
6. The tester can stop a test which he considers to be unsafe.

In these instances any fees paid in advance must be returned. However, if *during* the test the tester decides that it is unsafe to continue the test, then a fail

certificate will be issued and the reasons noted on it; and the full fee will be payable. For example, a petrol leak which was not evident at the start could be a reason for not completing a test.

DOCUMENTATION

What you need for the test
It is not normally necessary to have any documentation when presenting a vehicle for an MOT test, but sometimes the tester will need information contained on the registration document, so it is advisable to have it available. Information may be needed about the weight of commercial vehicles, and the vehicle age is often required.
   Many of the regulations depend on when the vehicle was first registered, due to the nature of the *Construction and Use Regulations* in force at the time. On the other hand, the testing station may not be authorized to test commercial vehicles above a certain weight, and can refuse to test a vehicle if it cannot be shown that it is within the weight limit of the testing

Department of Transport
Road Traffic Act 1988 sections 45 and 46

Customer's Copy

Serial No. 0 0842492 BB

Notification of a Refusal to Issue an MOT Test Certificate
(IMPORTANT: Read notes overleaf)

● *Please keep this copy as your record of the test failure.*

Vehicle Reg. Mark: C519 ATB Make & Model: Rover 1.6 Weight: — Year of Manufacture: 1986 Class of Vehicle: I II III IV V VII (IV ticked)

VIN or Chassis No.: SAXX82LY9AD18416 Recorded Mileage: 83428 Colour: BLUE

A Items tested.	Manual ref. Class V	Cars & LGV	Motor Cycles	Pass	Fail	Reasons for failure
Lighting Equipment	Class V	Cars & LGV	Motor Cycles			
Front & rear lamps etc	I/1,6,8	1.1	I/1	–		
Headlamps	I/2	1.2	I/2	–		
Headlamp aim	I/9	1.6	I/6	–		
Stop lamps	I/3	1.3	I/3	–		
Rear reflectors	I/4	1.4	I/4	–		
Direction indicators & hazard lamps	I/5,7	1.5	I/5	–		
Steering and Suspension	Class V	Cars & LGV	Motor Cycles			
Steering control	II/1	2.1	II/1	–		
Steering mechanism/system	II/2	2.2	II/2	–		
Power steering	II/3	2.3	N/A	N	A	
Transmission shafts	II/5	2.5	N/A	–		
Wheel bearings	II/5	2.5	II/3,4	–		
Front suspension	II/4,5	2.4,5	II/3	–		
Rear suspension	II/4	2.4,6	II/4	–		
Shock absorbers	II/6	2.7	II/3,4	–		
Wheel alignment	N/A	N/A	II/5	N	A	
Brakes	Class V	Cars & LGV	Motor Cycles			
ABS warning system/controls	III/3	3.4	III/1	–		
Condition of service brake system	III/3,5	3.3,5,6	III/2		X	Brake Fluid leaking between Master cylinder & Servo Unit
Condition of parking brake system	III/1,2,4	3.1,2,5	N/A	–		
Service brake performance	III/7	3.7	III/3	–		
Parking brake performance	III/7	3.7	N/A	–		
Additional braking devices	III/6	N/A	N/A	–		
Tyres and Wheels	Class V	Cars & LGV	Motor Cycles			
Tyre type	IV/1	4.1	IV/1	–		
Tyre load/speed ratings (Class V & VII)	IV/1	4.1	N/A	–		
Tyre condition	IV/1	4.1	IV/1	–		
Roadwheels	IV/2	4.2	IV/2	–		
Seatbelts	Class V	Cars & LGV	Motor Cycles			
Mountings/Condition/Operation	VI/1	5.1	N/A	–		
Motor Cycle Sidecar	Class V	Cars & LGV	Motor Cycles			
Security	N/A	N/A	V/1	N/A		
Suspension & wheel bearings	N/A	N/A	V/2			
Wheel alignment	N/A	N/A	V/3			
General	Class V	Cars & LGV	Motor Cycles			
Drivers view of the road	VI/4,15	6.1	N/A	–		
Horn	VI/5	6.2	VI/1	–		
Exhaust system	VI/8	6.3	VI/2		X	Exhaust leaking at flange between engine and exhaust
Exhaust emissions	VI/10	6.4	N/A	–		
General vehicle condition	VI/1,2 11,12,13 14,16,17	6.5	VI/3		X	o/s front chassis Corroded to excess
Mirrors	VI/3	6.6	N/A		X	o/s door mirror missing.
Fuel system	VI/9	6.7	N/A	–		
Registration plates and VIN numbers	N/A	6.8	N/A	–		
Speedo	VI/6	N/A	N/A	–		
Speed limiter	VI/7	N/A	N/A	N	A.	

SPECIMEN

B I Certify that for the reasons shown above the vehicle has failed to comply with the statutory requirements

Signed: (Tester/Inspector) A.N. Other Date: 30-3-93.

Name: (in CAPITALS) A.N. OTHER Testing Station No.: 14979

Authentication Stamp

C Warning: In my opinion, the vehicle is dangerous to drive because of the following defects*

See Brake Section

*delete wording if not appropriate

VT30 (Rev. March 1993)

Figure 9.1. *The test failure certificate must itemize each failure item. Note the comment in the danger box concerning the leakage of brake fluid.* (Crown Copyright Reserved. Reproduced with the permission of the Controller of Her Majesty's Stationery Office.)

station's authorization.

In these cases the vehicle may have its weight displayed on the plate containing the chassis and engine serial numbers, so it is well worth checking. Alternatively, as a rule of thumb, if a light commerical vehicle has twin wheels on the rear, its weight could put it outside the authorization of 'normal' car testing stations.

Older vehicles are exempt from various aspects of the test, but if the age cannot be proved they have to be tested normally, and you run the risk of being failed on testable items which should not really apply. In fact, the tester can refuse to conduct a test if he is not allowed sight of the registration document to enable him to find out what age- or weight-related test procedures are applicable to the vehicle in question.

The pass and fail certificates
When a vehicle is tested, it must either pass or fail. The tester *must* decide one way or the other on each testable item. If any one item fails, then you will be issued with an MOT fail certificate. This is officially labelled a Notification of Refusal to Issue an MOT Test Certificate (document VT30).

The fail certificate: Until recently the test failure sheet (VT30) (see Figure 9.1, which is a sample only) was a check-list upon which not only failed items were marked, but on which the tester could comment if an item only just passed, and was in need of early replacement or if during the test he had noticed a fault in a non-testable item worth noting. This is no longer the case, the Vehicle Inspectorate now insist that this form can only be used to note testable items which have failed the MOT.

However, most good testing stations now print their own check-list (see Figure 9.2) which clearly notes both items which only just pass, and other items, not part of the MOT test, which the tester may notice as requiring attention. It must be stressed that such documentation no longer carries any official status, but is of value to the person having the car tested. If you buy a car which has a recent MOT certificate, it is worth asking the garage which did the test if such a document was issued with the certificate, and if there was anything on it of interest!

The pass certificate: The pass certificate will show details of the car which was tested. It will have a date of issue, and a date of expiry. It should carry the signature of the tester who conducted the test, and an embossed name and address of the testing station (see Figure 9.3).

Duplicate certificates: If a certificate is lost, then up to 18

Figure 9.2. Most good test stations will produce their own checklist for their customers on which are listed items which have only marginally passed the MOT, or non-test items which need attention and which have been noticed during the test.

PUNTERS GARAGES
★ ★ ★ ★

FRIENDLY FAMILY FIRM
FAST EFFICIENT SERVICE
081-561 4327   081-573 1750
22 New Road, Hillingdon, Middlesex

M.O.T. TEST: ADVISORY NOTES

Although your car has passed/failed the M.O.T., we would advise you that the following items may require attention:-

LIGHTING EQUIPMENT	
STEERING AND SUSPENSION	
BRAKES	
TYRES AND WHEELS	
SEAT BELTS	
M/CYCLE SIDECAR	
GENERAL	

WARNING: IN OUR OPINION THE VEHICL FOLLOWING DEFECTS:-

REPAIR ESTIMATE:
    LABOUR: £

| PASS No. | REGISTRATION No. |
| FAILURE No. | |

Figure 9.3. The final goal – an MOT pass certificate. It can be forward dated up to a calendar month to the same month date of the old certificate provided the old certificate is available.
(Crown Copyright Reserved. Reproduced with the permission of the Controller of Her Majesty's Stationery Office.)

Keep this Certificate safely   See notes overleaf

The Department of Transport

Test Certificate

Serial number

The motor vehicle of which the Registration Mark **H 797 JJB**   **OY 0289544**

having been examined under section 45 of the Road Traffic Act 1988, it is hereby certified that at the date of the examination thereof the statutory requirements prescribed by Regulations made under the said section 45 were complied with in relation to the vehicle.

Vehicle identification or chassis number	WDB2010242F740969		
Vehicle Testing Station Number	14979	Vehicle colour	SILVER
Date of issue	Aug. 13 1992	Vehicle make	MERCEDES
		Approximate year of first use	1990
Date of expiry	Aug. 12 1993 Three	Recorded mileage	57341
		If a goods vehicle, max design gross weight	kg
Serial Number of immediately preceding Test Certificate	(To be entered when the above date of expiry is more than 12 months after the above date of issue.)	If not a goods vehicle, horse power or cylinder capacity of engine in cubic centimetres	2000
		Fuel type	petrol

Signature of tester/inspector

A.J. Smith

Name in BLOCK CAPITALS   **A.J. Smith.**

WARNING   Authentication Stamp

WARNING
A Test Certificate should not be accepted as evidence of the satisfactory mechanical condition of a used vehicle offered for sale.

CHECK
carefully that the particulars quoted above are correct. Certificates showing alterations should not be issued or accepted. They may delay the renewal of a Licence.

VT20

8394234.220037.1/93

months after the test was carried out a duplicate can be issued by the testing station, or by the district office of the Vehicle Inspectorate if the station no longer exists, or no longer has authority to carry out MOT tests.

Forward-dating of pass certificates: If a vehicle is tested and there is an existing pass certificate in force which expires inside a calendar month into the future, then provided this is made available to the testing station a new pass certificate can be forward-dated to expire a year hence from the same calendar date as the old certificate. This is useful if the motorist wants to have the car tested early so that repairs can be effected before the certificate expires. It means that he or she is not penalized by losing time from the existing certificate.

THE RE-TEST

If a car fails the MOT test (apart from a few exceptions noted below) it will have to be *fully tested* again when it is brought back to the testing station for re-test. This means that if, in the meantime, some other testable item has failed, then another test failure will result.

The need for a full test on re-presentation can be a difficult aspect of the regulations from the point of view of the testing station. Customers do not understand why they have to endure the process all over again when all they have done (for example) is to go to the local tyre company to have a tyre fitted. Nevertheless, that is the rule laid down by the testing authorities and it has to be complied with. Their attitude has always been that on even the shortest journey a testable item on a car can fail, which on resubmission would prevent a pass certificate being issued. This is no longer so easy to accept since they have now applied some exceptions to the rule.

In the following cases, if the vehicle is taken back to the testing station before the end of the next working day then these items can be checked without a full re-test and without any further fee: Horn, direction indicators/hazard warning lights, headlamp aim, lamps, rear reflectors, registration plates, VIN, mirrors, seat belts (excluding anchorages), doors, bootlids, tailgates, loading doors, tailboards, dropsides, windscreen, washers/wipers, exhaust emissions.

A surprising omission here is the petrol cap. Although this is simplicity itself to replace and then inspect, the Vehicle Inspectorate insist that as part of the fuel system the whole vehicle must be inspected again.

Figure 9.4. *MOT test stations must provide customers with a Notice of Appeal if requested.* (Crown Copyright Reserved. Reproduced with the permission of the Controller of Her Majesty's Stationery Office.)

Notice of Appeal

Road Traffic Act 1988 Section 45

COMPLETE IN CAPITAL LETTERS ✓ *the boxes*

Please read the notes on page 4.

VT17
Rev Feb/92

I/We would like to appeal to the Secretary of State for Transport against the refusal of a Test Certificate for the vehicle described below:

1. Name of Appellant
2. Address
   Postcode
3. Daytime telephone number
4. Name of Vehicle Testing Station involved
   Address
   Postcode
   Testing Station No. *(see VT30)*
5. Give the reasons for making your appeal
6. Make and Model
7. Vehicle Registration No.
8. Chassis No.
9. Engine Capacity
10. Colour
11. Year of Manufacture
12. Type of Vehicle: Motorcycle / Motor car / Commercial Vehicle / Motorcycle and sidecar / Three wheeled motor car
13. If it is a commercial vehicle, please give unladen or gross weight
    Unladen weight kgs Gross weight kgs

THE RIGHT TO APPEAL

If a motorist is not satisfied that the test on his vehicle has been done properly there is an appeal procedure wherein the Vehicle Inspectorate consider the situation and re-test the vehicle themselves. All testing stations must have Notice of Appeal forms (document VT 17) available to enable customers to take advantage of this procedure if they so desire (see Figure 9.4).

All appeals must be made within 14 working days from the date of the test, and the complainant must not adjust or repair the car prior to examination by the Vehicle Inspectorate, and be informed to this effect by the testing station. A further full fee applies irrespective of the outcome of the appeal.

SPECIAL TEMPORARY CONCESSION

It is a condition of driving a vehicle on the road that it has a valid MOT certificate if one is required. In fact, it is not possible to obtain a road fund licence without a current MOT certificate. To avoid the obvious problem of drivers illegally taking vehicles for test which do not have a current certificate, the regulations allow drivers to go to and from an MOT station for a test without road tax, or an MOT. (But remember, the vehicle must still be insured, and problems which could render the car unroadworthy may invalidate the insurance.)

To do this the testing station must be notified and provide the earliest practicable appointment for the test. When they have recorded the vehicle details, the name of the person and the appointment time and date, then the vehicle can be legally driven to and from the testing station for the appointed test. This

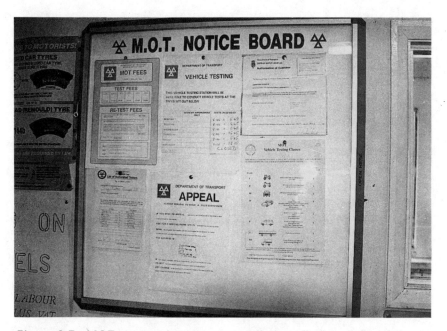

Figure 9.5. *MOT test stations are compelled to display some of the information connected with the MOT test.*

concession applies only to road tax and MOT regulations and not to the requirement that a vehicle should be roadworthy. If the car had bald tyres, for example, a successful prosecution could still follow under the Construction and Use Regulations if the police decided to proceed.

THE COST OF THE MOT

The price of an MOT test is reviewed every year by a working committee consisting of representatives from the testing authority and bodies who represent the motor trade. It is adjusted according to inflation, and reviewed if new regulations mean that additional time is required to undertake a test.

Over the years the price has steadily risen, and quite sharply in recent times mainly due to the additional equipment and extra work required to complete a test. This is not surprising with the emphasis moving more and more to road safety and environmental factors. The focus is now shifting to the quality and qualifications of people who want to become

MOT testers, which inevitably means that the cost will continue to rise as testing stations have to employ more highly qualified personnel to test vehicles.

Rapid check

Simple guide and checklist to the MoT (for cars and light commercial vehicles)

LIGHTS

Sidelights:
The fronts must be white, or yellow if they are inside a tinted headlamp, and the rears must be red. A broken lens is a failure unless it has been properly repaired with special tape. Make sure they work without flickering even if tapped, and that the switch works properly.

CHECK LIST
* **Are they the correct colour?** ☐
* **Do they all work without flickering?** ☐
* **Does the switch work properly?** ☐

Headlamps:
Headlamps can be single or double units and different cars have different lens designs. Make sure they all work on both main and dip, and that the glass is not broken. Check the condition of the reflectors and that the bulbs are fitted correctly.

You cannot check the aim or beam shape without proper beam testing equipment but if the beam is shone onto a wall, the dipping action can be checked, and you can make sure that the beams are the same shape each side, not fuzzy, and that the aim is not much too high or low.

Finally, make sure the lamps are not loose, there are no poor connections causing flicker and that the switch works properly.

CHECK LIST
* **Do all the headlamp beams work on main and dipped?** ☐
* **Do they flicker when tapped?** ☐
* **Is the glass broken?** ☐
* **Is the reflector corroded?** ☐
* **Are the beam shapes about right?** ☐
* **Is the aim much too high or low?** ☐
* **Do the on-off and dip switches work correctly?** ☐

Stop lamps:
Stop lamps must come on when the brake is applied, and go off when it is released. If 'sticky' either way, they fail. Make sure the lenses are not broken, and there is no flicker when tapped.

CHECK LIST
* **Do the stop lights work and come on and off with the brakes?** ☐
* **Are the lenses unbroken?** ☐

Flashing indicators:
Indicators must flash between 60 and 120 times a minute. If too slow, rev the engine; this may solve the problem. Cars first used after 1 September 1965 must have amber indicators. On those used before, the fronts could be white and the rears red. Side repeater indicators also have to be tested if the car was first used on or after 1 April 1988.

Make sure that the switch works properly, as does the 'tell-tale' light or audible signal inside the car. The self-cancelling mechanism is not part of the test.

CHECK LIST
* **Do the required indicators work at between 60 and 120 times per minute?** ☐
* **Are the lenses the correct colour and unbroken?** ☐
* **Does the switch work properly?** ☐
* **Is the 'tell-tale' working?** ☐

Earthing faults:
A common failure on lights is an earthing fault. This shows up when, with the side lights on, operating the brake lights or indicators causes one or more of the lights to malfunction. This is often caused by poor earthing in the light unit, often as a result of corrosion.

CHECK LIST
* **With the side lights on do stop lights and indicators all work correctly?** ☐

Reflectors:
On most cars they will be an integral part of the rear lamp cluster. Make sure that they are clean and not broken and that no badges and so on obscure them.

CHECK LIST
* **Are the reflectors clean, unbroken, and unobscured?** ☐

STEERING & SUSPENSION

Steering wheel & column:
Test the wheel for strength and condition by pushing and pulling along the shaft axis, and at right angles to it. This also indicates wear between the shaft and the steering wheel, showing if the wheel is not properly attached or that the top bearing is worn. It will show up excessive steering shaft end

float, and check the upper shaft mounting bracket.

Finally, rotate the wheel both ways to check steering column couplings and clamping bolts, but do not confuse movement with play in the steering box or rack and pinion which is checked separately.

CHECK LIST
* **Is the steering wheel unbroken and are the mountings secure?** ☐
* **Are the steering column bearings excessively worn?** ☐
* **Are the column couplings excessively worn?** ☐
* **Is there excessive 'end float'?** ☐

Steering mechanisms:
Check the play in the steering box or rack and pinion by waggling the steering wheel from side to side until the front wheels just move. Up to 13mm of free play at the steering wheel is allowable for rack and pinion systems, and up to 75mm if a steering box is fitted.

The rest of the steering mechanism is checked by turning the wheel one way and another until there is resistance. Then check for wear in all the different steering swivel joints. Make sure that all the locking devices are in place (split pins, locking nuts etc.) and that there are no tears or splits in the steering rack gaiters. On the few cars with some rear wheel steering, all the links, bushes and joints will have to be checked.

Check that the steering box or rack and pinion is not loose on its mountings, and that they are not cracked. Have a good look around the steering mechanism to check there are no obvious problems when the wheels are turned from lock to lock. With power steering, run the engine during these checks, and look

for leaks and check that the pump and pipes are in reasonable condition.

Finally examine the surrounding chassis areas and make sure there is no corrosion, distortion or cracking in the structure within 30cm of any steering mechanism attachment points.

CHECK LIST
* **Are all the steering joints in reasonable condition?** ☐
* **Is the steering box or rack and pinion worn excessively?** ☐
* **Are the rack gaiters torn?** ☐
* **Are there any split pins or locking nuts/washers missing?** ☐
* **Do the wheels turn lock to lock without fouling?** ☐
* **Does the power steering work properly without leaks?** ☐

Front and rear suspension:
All the joints on the suspension arms, and the main steering/suspension swivel joints must be checked for wear. The vehicle must be jacked up to relieve the load off the steering/suspension swivel joints so they can be properly examined.

Check the springs to make sure they are not fatigued or broken. Shock absorbers must be securely attached, leak-free and the bushes inspected. Bounce the car up and down to see if the damping effect works properly.

Have a good look around and make sure that all the split pins, locking nuts and so on are where they should be. Also check that the anti-roll bar is secure and that its bushes are acceptable. On cars with MacPherson strut suspension, check the upper swivel bush under the bonnet. On vehicles with Hydrolastic suspension, check for leaks and make sure

the car sits level. While the wheels are off the ground, the wheel bearings and drive shaft can be inspected for excessive wear or play. If the car has CV joints these must be checked as far as possible, and the CV boots must not be torn or split.

CHECK LIST
* **Has the car been jacked to relieve the load from the joints?** ☐
* **Are all the suspension/steering joints and bushes OK?** ☐
* **Are the road springs fatigued or broken? Does the Hydrolastic leak?** ☐
* **Do the shock absorbers work properly and not leak?** ☐
* **Are the wheel bearings and drive shaft systems OK?** ☐
* **Are the CV boots torn?** ☐

BRAKES

Without specialised equipment you cannot fully check the brakes but much can be done. Inside the car, see if the footbrake is firm and does not depress nearly to the floor. Operate the handbrake to see if the ratchet works and releases properly. Check the handbrake mounting for cracks or corrosion. With power-assisted brakes, the servo must be working. If the car has ABS, the warning light must be checked to make sure it illuminates and follows the correct sequence for that car.

Under the bonnet examine the brake master cylinder for leaks and see if it is full of fluid. Check pipes and unions for condition and leaks. Then, beneath the car, look at all the flexible brake hoses for condition and make sure they are not chafed by the wheels on full lock. Also examine all the metal pipes and unions for leaks and corrosion, together with any compensating valves which are sometimes fitted.

Have a good look at the disc pads if they are visible, and the brake calliper. On vehicles fitted with drum brakes, check the back plates for signs of fluid leakage from the wheel cylinders inside. With an assistant operating the handbrake, have a look at the cables and mechanism to check that it is all working properly and none of the levers or cables are seized.

The best that can be done to check brake performance is to drive on a deserted road at about 20mph, gradually apply the footbrake and check that the car does not veer one way or the other, that there is no brake squeal, or 'rumble' from the pedal. Then do the same with the handbrake. Remember, even if the brakes seem to be OK, you still cannot be sure until they are checked at the MOT station.

Thoroughly check the road ahead and behind as a safety measure first.

CHECK LIST
* **Is the footbrake firm and reasonably resistant?** ☐
* **Does the handbrake ratchet lock properly?** ☐
* **Is there reserve movement on the handbrake?** ☐
* **Is the brake servo working?** ☐
* **Is the brake fluid reservoir full? Are there any fluid leaks anywhere in the system?** ☐
* **Are the flexible hoses chafed or the metal pipes corroded?** ☐
* **Are the brake pads worn too low?** ☐
* **Does the ABS warning light work properly?** ☐

* **Is the handbrake mechanism freely operating?** ☐
* **Do the brakes work properly when operated to stop the car?** ☐

EXHAUST SYSTEM

From under the car, check the exhaust system. Make sure it is fitted securely from the engine exhaust manifold down to the final mounting at the tail pipe. With the engine running, check for leaks throughout the system. Lastly, make sure that it is not too noisy. This can happen when internal baffles fail, even if the outside seems OK. Any leak means a failure but a good repair with paste and/or tape is acceptable.

CHECK LIST
* **Does the exhaust leak?** ☐
* **Is it properly attached?** ☐
* **Is it too noisy?** ☐

EXHAUST EMISSIONS (petrol)

This was only introduced in 1992 and applies to cars first used on or after 1 August 1975. The DIY motorist cannot check this.

The test is threefold. First, a failure will result if the vehicle emits dense blue or black smoke for more than five seconds at idle after running the engine to about 2,500rpm (or about half throttle) for 20 seconds.

Secondly, the hydrocarbon content is checked and must not exceed 1200 parts per million for a continuous period of five seconds.

Thirdly, the carbon monoxide is checked. Two levels apply depending on the age of the vehicle. Those first used before 1 August 1983 must not have a carbon monoxide content in the exhaust gases exceeding 6% for a period of five seconds. All later

cars must not exceed 4.5% for the five-second period.

CHECK LIST
* **You can check only if the car emits dense smoke.** ☐

EXHAUST EMISSIONS (diesel)

Special equipment is used for the test so the average motorist will be unable to check for himself.

CHECK LIST
* **All that can be done is to rev the engine to the speed limiter and see if there is excessive smoke. In the test this has to be done a number of times.** ☐

STRUCTURAL SECURITY & CORROSION

A car can fail the MOT test as a result of corrosion in two ways. First, if structural areas of the vehicle are damaged, distorted within 30cm of a brake, suspension or steering component.

Structural areas of the car are specifically prescribed in *The MOT Inspection Manual* and are shown in detail in Chapter 3, on pages 26 and 27.

Excessive corrosion is checked by squeezing the suspect metal or component between the finger and thumb to see if it 'gives' or crumbles. If so, then it must fail. Light tapping or scraping with a special tool, which is in effect a small 'toffee' hammer, is acceptable to detect corrosion. If light tapping penetrates the surface, or scraping reveals underlying corrosion which has weakened the structure, a failure results.

Repairs have to be welded and carried out in a special way depending on the nature of the structure being repaired. This is laid out in detail in the Manual. Highly-stressed components (a suspension arm for example), generally cannot be repaired by welding and will have to be replaced.

CHECK LIST
* **Is the main load bearing structure corroded to excess?** ☐
* **Is there excessive corrosion within 30cm of a brake, suspension or steering component?** ☐
* **Have weld repairs been properly carried out?** ☐

TYRES & WHEELS

The tyres fitted to any one axle must be the same type and size (ie. the two fronts must be the same, and so must the rears). If cross-ply or bias-belted tyres are fitted to the front, then a failure results if radial-ply tyres are fitted to the rear. A failure also results if cross-plies are fitted to the front, and bias-belted tyres to the rear. Make sure none of the tyres fouls any other parts of the car. The spare tyre is not tested for the MOT.

A tyre must not have excessive cuts, lumps or bulges, or have had its tread re-cut. It must be correctly seated on the rim. The valve must be in good condition and properly aligned. A worn tyre will fail the MOT. There must be tread to at least 1.6mm depth all around the tyre for the centre three-quarter of the tyre width.

An under-inflated tyre is not a failure but it could affect the brake performance and headlamp alignment aspects of the test.

Finally, the road wheels must not be damaged. This means that if the wheel has been

'kerbed' it must not have damaged the rim or distorted the wheel so it runs out of true. Check that none of the wheel nuts or studs are missing.

CHECK LIST
* **Does the car have the right combination of tyre type fitted?** ☐
* **Do any of the tyres have splits or bulges which would result in a failure?** ☐
* **Are any of the tyres worn beyond the legal limit?** ☐
* **Are the road wheels damaged or distorted?** ☐
* **Make sure none of the wheel nuts or studs are missing.** ☐

SEAT BELTS

The car's age is important here. Those first used before 1 January 1965 do not need seat belts. On cars first used on the road from then until 31 March 1987 front seat belts are required, and all later cars must have both front and rear belts tested. In fact things are a bit more complex than that, as noted on pages 92 and 93.

Make sure the belts are in good condition, operate correctly, that the mounting points are secure and there is no corrosion or structural damage within 30cm of the mountings.

CHECK LIST
* **Does the car need to have rear seat belts?** ☐
* **Are the belts in good condition?** ☐
* **Are the mountings and attachment points secure with no corrosion or structural damage within 30cm?** ☐

WIPERS, WASHERS & HORN

Wipers and washers are not needed if the windscreen can be opened to give the driver a good view of the road. Otherwise the wipers must work and give the driver a good view of the road to both sides and to the front. The wiper blades must be in good condition, with no peeling or tears in the rubber. They must be properly attached and secure. Finally, the switch must work correctly. The washers need only operate to provide enough liquid to clear the windscreen adequately when the wiper operates.

CHECK LIST
* **Do the wipers work properly?** ☐
* **Do the washers work properly?** ☐
* **Does the operating switch work correctly?** ☐
* **Are the wiper blades in good condition?** ☐

The horn must be loud enough to be heard. Sequence multi-tone horns will fail. Two-tone horns which operate together are acceptable. The switch must work properly and be easily reached by the driver.

CHECK LIST
* **Does the horn work when the switch is operated?** ☐
* **Does the horn have the legal tone?** ☐
* **Can the switch be easily reached by the driver?** ☐

FOG LAMPS

On cars first used on or after 1 April 1980 a rear fog lamp must be fitted to the rear, in the centre or offside. This now has to be tested.

Make sure it is red, operates

properly and does not have a broken lens. There must be a 'tell tale' which the driver can see. Its operation must be consistent and unaffected by other lights.

CHECK LIST
* **Does the car need a rear fog lamp?** ☐
* **Is it positioned properly?** ☐
* **Is the lens broken?** ☐
* **Does the switch work properly and illuminate the 'tell tale' light?** ☐
* **Does it work properly without flickering, or being adversely affected when other lamps are switched on?** ☐

NUMBER PLATE LAMPS

If the car has one number plate lamp it must be working. If it is meant to have more than one lamp, then if any one lamp or bulb is not working a failure will result. Make sure that the required lights work without flicker.

CHECK LIST
* **Do all the number plate bulbs and lamps work?** ☐
* **Do they flicker?** ☐

HAZARD WARNING LIGHTS

Hazard warning lights are now tested in the MOT. This applies to any cars fitted with hazard warning lights, and for all cars first used after 1 April 1986.

Regarding condition of the lenses and operation, the situation is the same as for the rest of the indicators. The hazards must all flash in phase and have a working 'tell tale' inside the car.

CHECK LIST
* **Carry out the same checks as for the indicators. Do all the lights flash together?** ☐
* **Does the in-car 'tell tale' work?** ☐

WINDSCREEN

The windscreen must not be damaged or have obstructions which obscure the driver's view of the road.

The damage is defined within two 'zones' on the screen. The first zone is a strip 290mm wide more or less in front of the driver, vertically up to the limit of the 'wipe' of the wipers. In this zone any 'damage' or obstruction must be contained inside a 10mm circle. The second zone is the rest of the 'wipe' of the wiper blades. Damage or obstruction in the second zone must fall inside a 40mm diameter circle.

Any aspects of the original car design which impinge on these limits will not result in a failure. Official stickers are allowed.

CHECK LIST
* **A split from top to bottom on the driver's side within the wiped area of the driver's wiper blade will be a failure.** ☐
* **Do any other chips or damaged sections within the two zones extend beyond the 10mm and 40mm circular limits?** ☐

BODYWORK CONDITION

If bodywork is damaged or corroded so that there are sharp edges which could cause injury, then a failure will result.

On vehicles which have a separate chassis and body, the security of the attachment of the body to the chassis is now tested.

CHECK LIST
* **Check for sharp edges from corrosion or damage which could cause injury.** ☐
* **With separate body/chassis check attachment points.** ☐

DOORS, BOOTLIDS, TAILGATES ETC.

The front doors must open from both inside and outside the car. The rear doors only have to latch shut.

Bootlids, tailgates, loading doors, dropsides and so on must be able to be secured in the closed position, but any original design aspects which fall foul of this can be accepted.

CHECK LIST
* **Do the front doors operate from both inside and outside the car?** ☐
* **Do all the other doors, bootlids, tailgates etc. close and lock properly?** ☐

DRIVER & FRONT PASSENGER SEATS

Both these seats are checked to ensure they are not loose or broken, and that the back rests correctly lock.

CHECK LIST
* **Make sure that the seat slider locks properly, as does the seat back.** ☐
* **Check that the seat frame is not broken.** ☐

REAR VIEW MIRROR(S)

If the car was first used before

1 August 1978 it need have only one rear view mirror. Later cars must have two mirrors, one of which must be on the driver's side.

CHECK LIST
* **Does the car have enough mirrors for its age?** ☐
* **Are they damaged or obscured so the rear view is seriously impaired?** ☐
* **Can the driver see them from the driving position?** ☐

FUEL SYSTEM

The fuel system from the fuel tank to the engine must be checked both with and without the engine running. There must be no leaks in any of the hoses, pipes and unions or from the engine. The fuel tank cap must be secure and not likely to leak.

CHECK LIST
* **Is there any fuel leak throughout the fuel system?** ☐
* **Are any of the pipes/hoses/unions in a poor condition or insecure?** ☐
* **Does the tank cap fit properly and seal the tank?** ☐

A temporary/emergency fuel cap will fail.

NUMBER PLATES

All vehicles must have both front and rear number plates. These must not be broken or have numbers/letters missing. The layout of the letters and numbers must be conventional, and the number plates must not be obscured.

Vehicles used on or after 1 August 1980 must have a plainly displayed and legible Vehicle Identification Number (VIN).

CHECK LIST
* **Is the number plate broken or has it got letters missing?** ☐
* **Are the letters/numbers properly displayed?** ☐
* **If required, does the vehicle have a clear VIN number displayed?** ☐